The
Hempcrete
Book

The Hempcrete Book

Designing and building with hemp-lime

William Stanwix and Alex Sparrow

Published by
Green Books
an imprint of UIT Cambridge Ltd
www.greenbooks.co.uk

PO Box 145, Cambridge CB4 1GQ, England
+44 (0)1223 302 041

First published 2014, in England. Reprinted 2016, 2018, 2021

Front cover photograph © Alex Sparrow
Back cover photographs: left © Alex Sparrow; middle © Adnams; right © Jørn Tomter

Design by Jayne Jones

All interior photographs © Alex Sparrow, except where otherwise credited. Photographs of
Hemp-LimeConstruct builds on pages 8, 24, 128-9, 130, 230, 244, 247, 249, 288 and 317 are
© Jørn Tomter (www.tomter.net) and on pages 31 (right) and 35 are © Enrique Neyland Quintano.

ISBN: 9780857842244 (hardback)
ISBN: 9780857841209 (paperback)
ISBN: 9780857841223 (ePub) ISBN:
9780857841216 (PDF for libraries)
Also available for Kindle

Disclaimer: The advice herein is believed to be correct at
the time of printing, but the authors and publisher accept
no liability for actions inspired by this book.

ep-1-4

Contents

Foreword by Professor Tom Woolley.............. 7

Introduction 9

Part One: Principles of building with hempcrete

1 History and uses of hemp........................... 15
2 Hemp in construction 23
3 An introduction to lime 41
4 Key concepts in sustainable building...... 53
5 Getting the basics right 67
6 Variations on the hemp–lime mix 79
7 Performance of hempcrete
   in a building.................................... 87
8 Tools and equipment 101
9 Health and safety............................... 109
10 Planning the build 117

Focus on self-build 1:
Agan Chy 124

Part Two: Hempcrete construction

11 The hempcrete wall: an overview 131
12 Foundations and plinth.......................... 147
13 The structural frame 155

14 Shuttering.. 169
15 Mixing hempcrete 191
16 Placing hempcrete................................. 207
17 Floors, ceilings and roof insulation........ 221
18 Finishes for hempcrete 231
19 Practicalities on a hempcrete build....... 253
20 Restoration and retrofit.......................... 267

Focus on self-build 2:
Hemp Lime House.. 282

Part Three: Designing a hempcrete building

21 Design fundamentals 289
22 Indicative detailing................................ 309

Focus on self-build 3:
Bridge End Cottage 342

23 A look to the future................................ 347

Notes 350
Glossary 352
Resources... 354
Conversion tables 358
Bibliography 359
Index... 360

This book is for
Jacob John Renbourn

Not only a talented builder but also a true
hemp devotee. He is greatly missed.

Acknowledgements

First and foremost, huge gratitude goes to our wives and children for their patience and understanding throughout the writing of this book (and the rest of the time too!).

We are grateful to the folks at the Low-Impact Living Initiative (LILI); especially to Dave Darby and Elaine Koster for the original spark which ignited this project and Emma Winfield Tubb for her hard work both at LILI and on-site, and her friendship and support.

We are indebted to the team at UIT / Green Books, past and present, for giving us the opportunity to write this book, and to Alethea Doran and Jayne Jones, for their hard work and patient guidance, which have allowed it to become what it is today.

We'd like to say a special thank you to Tom Woolley of Rachel Bevan Architects, and Graham Durrant of the Limecrete Company, who read early drafts of the manuscript and provided helpful, and often entertaining, comments and advice.

The following have given generously of their time, expertise and advice during the writing of this book. Mark Patten and Dave Mayle at Lime Technology, Myles Yallop and Graham Durrant at The Limecrete Company, Nigel Gervis at Tŷ-Mawr, Neal Holcroft, Nick Voase at K J Voase and Son, James Ayres at Lime Green, and Mark

Womersley at Womersley's. The advice and information they have provided has been invaluable. The responsibility for any misinterpretations or mistakes in the finished work lies solely with the authors.

We are indebted to Dave, David, Jules, Eve and everyone else at Scarthin Books in Cromford, where Alex ended up being 'writer in residence' for several weeks, and to Elaine and Mark Loydall for the loan of their cottage during the final stages of this project. A big thank you also to Pat Gilgrass, who not only lent us her caravan during the editing stage, but heated it too!

We are grateful to Hemp-LimeConstruct's clients, past and future, without whom none of this would be possible; with special thanks going to Anthony and Penelope, Sarah and Tony, and Alexei and Sarah.

Last but by no means least, our thanks go to our team: Rob Mohr, Pam Stanwix, Riq Neyland Quintano and Aniko Hegedus; and to the Hemp-LimeConstruct extended family: Emma Winfield Tubb, Paul Fitzpatrick, Alaric 'Mighty' Federlein, Sam Thomson, Kris Clarke, Kate Noble, Heather Miles, and everyone else who has worked or volunteered for us over the last six years.

Massive thanks to www.radiomeuh.com for the background noise.

Foreword

When Rachel Bevan and I did the research that led to the publication of *Hemp Lime Construction* in 2006-7, we had a fairly good idea of the location of every building containing hempcrete in the UK and Ireland. Seven years or so later, it is impossible to keep track of the use of this remarkable composite building material, as happily its use has become almost commonplace in the UK – evidence of the widespread accept-ance of this excellent sustainable way of building. However, whenever such an innovative form of construction quickly gains popularity there is a risk, if it is used by people who expect it to behave like 'conventional' materials, of careless and poorly supervised construction, detailing and specification. For this reason *The Hempcrete Book* is most welcome, as it brings up to date the knowledge and experience gained in recent years and provides clear guidance to those who want to apply hemp materials correctly in a range of applications.

As pressure builds to meet increasingly strenu-ous energy-efficiency targets, many weird and wonderful building techniques and materials have appeared on the market. These are often embraced with remarkable haste, even if barely tried and tested, particularly when backed by glossy brochures and exaggerated performance claims. Most of these systems involve plastic and petrochemical-based substances which can sometimes present fire hazards, emit toxic chemicals and pollute the planet when disposed of in landfill. They use fossil-fuel resources and produce significant carbon dioxide emissions, although ironically their use is intended to reduce such emissions! Many so-called low-energy or zero-energy buildings have been shown to consume more energy in their materials and construction ('embodied energy') than is saved in the lifetime of the building.

Hempcrete, on the other hand, not only has low embodied energy but also locks up CO_2 in the building fabric. It is fireproof, healthy, breathable, regulates humidity levels and is much more thermally efficient than comparable materials. All of this is explained in detail in this excellent book. Once you become aware of hempcrete's advantages it is hard to find another way of building that can meet the demands of sustain-able, healthy and energy-efficient construction so successfully. Furthermore, since the hemp plant can be grown in so many parts of the world, it can be used to insulate buildings in poorer developing regions as easily as in Western countries. Hemp also provides food, oil, clothing, paper and many other products in addition to construction materials.

The construction and use of buildings are responsible for at least 50 per cent of the CO_2 emissions from human activity. Therefore using materials with a much lower environmental impact is crucial to reducing environmental damage and resource consumption. Designing and building with hempcrete is a real demon-stration of a total commitment to 'saving the planet' and protecting the health and well-being of a building's occupants. It's an easy commitment to make, because hempcrete is affordable, great fun to build with and is truly sustainable. Thanks to this book, it will now be even easier to use.

Tom Woolley
County Down, April 2014

Introduction

This is a book about 'natural building'. An increasing number of people across the UK, and the world, are consciously deciding to use natural materials when constructing new buildings or restoring old ones. People usually choose natural materials because they want a building that is sustainable or 'low impact' in terms of eliminating or minimizing any lasting negative effect on the world in which we live. In practice this means reducing as far as possible the 'embodied energy' associated with the materials and methods used in the construction of the building; minimizing the amount of fossil fuels needed by the occupants to power, heat and cool the building through its lifetime; and minimizing toxic emissions and any other harm to human society or the natural world.

People also choose natural materials because they are increasingly aware of how such materials can not only maintain the structural fabric of the building well but also help to keep humans in good health. In contrast to many synthetic materials, natural materials contain no harmful chemicals. They are also vapour permeable (they allow moisture to pass through), which has significant implications for the health of both the building and its occupants.

There are many natural materials available for use in construction: timber, stone, earth, animal hair, straw, hemp, lime, reed and fired clay, to name just some. Many of these materials can be used in more than one way. They have a long pedigree of use over centuries, or even millennia,

by humans (and other animals!) to provide warmth and shelter. Today, with our growing awareness of the threats associated with over-dependence on fossil fuels, and our increasing understanding of the negative side-effects of synthetic building materials mass-produced by highly industrialized processes, has come a resurgence of interest in older and more natural materials and methods for construction.

While many techniques from our construction history are suitable for use today, others are less applicable in the current context and probably have limited use in the new buildings of the future. However, one positive effect of this resurgence of interest is that a number of new natural materials and techniques, based on old technologies but tailored to meet our future construction needs, are beginning to emerge. Hempcrete (a hemp–lime composite construction material) is one such new material. Comprising the chopped stalk of the industrial hemp plant mixed with a lime-based binder, hempcrete provides a natural, healthy, sustainable, local, low-embodied-energy building material that can truly claim to be *better than zero carbon*. Carbon dioxide taken up by the plant when it was growing is locked up in its woody fibres, and at the end of the building's life the hempcrete can be left to compost and be used as a soil additive rather than going into landfill. As a highly insulating material with significant thermal mass, hempcrete has excellent thermal performance within the structure of a building, and there is increasing evidence that it actually performs much better in real-life situations than is suggested by steady-state modelling.

Left: Hempcrete and lime plasters in a listed building.

Our company, Hemp-LimeConstruct, was set up when we decided to focus on furthering the use of this exciting new material in building projects across the UK. We have been building with hempcrete since 2008, on projects ranging from 'future-proof' houses to large community buildings and complete restorations of listed buildings, and provide a bespoke design-to-build service, consultancy and training. At the same time, other contractors across the UK have been using hempcrete on a much larger scale, in the construction of housing estates and commercial and industrial developments. As the profile of hempcrete within the public continues to grow, there is an increasing demand for these services. And as research into hempcrete builds, our understanding of this remarkable material is constantly developing. From the interest we at Hemp-LimeConstruct get at exhibitions and the frequency of enquiries we receive, it looks as though hempcrete has caught the nation's imagination and is here to stay.

When we first sat down to write this book, we intended to write a practical 'how-to' manual to introduce people to the method of building with hempcrete. The reason for this was that we would have found such a book very useful when we first started working with the material, and it didn't

Freshly mixed hempcrete.

exist. We hope that others wanting to build with hempcrete will benefit from our six years of experimentation and refining of techniques; learning from *our* mistakes rather than their own.

As the book took shape, we realized the importance of including discussion of some key topics and concepts that underpin the successful use of hempcrete, and we hope readers will find this background interesting and useful. Many people will want to flick straight to the practical details contained in Part Two, but we'd urge all our readers to take the time to ensure they understand the contextual information and important underlying concepts discussed in Part One. A firm understanding of these is essential for anyone wishing to become a successful hempcrete builder.

Hempcrete is especially attractive to self-builders and community groups, because of the relatively low-tech nature of the construction method. Also, owing to the fact that it's a relatively labour-intensive construction method, big savings can be made by providing your own labour. However, as our company and others have proved over recent years, hempcrete is also commercially viable as a construction material in a wide range of applications. Its cost is comparable to that of conventional construction methods, but if you factor in the true cost of the embodied carbon of conventional building materials, in terms of environmental damage, and consider the financial benefits of the energy savings that hempcrete delivers through the lifetime of the building, you could argue that it's actually a lot cheaper!

We hope that readers will be inspired and encouraged by this book to develop their own hempcrete projects – but first, an important caveat. It is beyond the scope of this book to show every possible way of using hempcrete, and in any case the industry as a whole is still exploring this. Also, it is important to remember that while the

Hempcrete housing at The Triangle in Swindon.

principles and information outlined in these pages are sound, the physical properties of hempcrete change slightly depending on which binder you are using. As will become clear, it is also a material (like many others) that has the potential for problems associated with a lack of understanding or poor workmanship on the part of the builder. For these reasons, it is important to remember that on any build it is the responsibility of the building designer to specify the exact materials and design construction details, and the responsibility of the builder to ensure a high standard of knowledge and skill in those who are working with the material on-site.

There is a great pleasure to be found in building with hempcrete, which comes not only from the extraordinary thermal performance achieved but also from the simplicity: both of the material itself and of the elements within a typical hempcrete construction. Hempcrete's low-tech nature means that, with relative ease, highly energy-efficient buildings can be constructed that contain virtually no synthetic, highly processed or high-embodied-carbon materials. With a good understanding of the material, and a little practice, hempcrete is a hugely rewarding material to work with, and can produce beautiful, healthy, 'future-proof' buildings.

PART ONE

Principles of building with hempcrete

CHAPTER ONE
History and uses of hemp   15

CHAPTER TWO
Hemp in construction   23

CHAPTER THREE
An introduction to lime   41

CHAPTER FOUR
Key concepts in sustainable building   53

CHAPTER FIVE
Getting the basics right   67

CHAPTER SIX
Variations on the hemp–lime mix   79

CHAPTER SEVEN
Performance of hempcrete in a building   87

CHAPTER EIGHT
Tools and equipment   101

CHAPTER NINE
Health and safety   109

CHAPTER TEN
Planning the build   117

CHAPTER ONE

History and uses of hemp

The hemp plant, thanks to its many uses and in particular its most famous one, as a widely popular recreational drug, is one of the most instantly recognizable plants in the world. A great deal has been written about hemp's many uses throughout human history and about the politics of its prohibition during the twentieth century, and there is no need for us to reproduce it all here. In the interests of context, however, in this chapter we provide a brief description of the hemp plant, its history and the resurgence in its use today.

Hemp is the English name (from the old English *haenep*) for the cannabis plant. The words *haenep* and cannabis are both thought to derive from the Ancient Greek *kánnabis*, which in turn evolved from an older word in an ancient Iranian language from around 2,500 years ago.

Three varieties of the cannabis plant exist: *Cannabis sativa*, *Cannabis indica* and *Cannabis ruderalis*. *Cannabis sativa* and *C. indica* are seen as the more closely related species. *Cannabis ruderalis* differs from them in that its flowering happens after a predetermined number of days, rather than being dependent on the seasons, and it contains very little tetrahydrocannabinol (THC), the psychoactive substance that gives the drug cannabis its active ingredient.

Hemp is a fast-growing erect annual plant which produces only a few branches, usually at the top of the plant, and grows to a height of between 1.5m and 4m. Its stem is thin and hollow, with a diameter of 4mm to 20mm, depending on the conditions and the specific variety grown. The 'bast' fibres of the hemp plant, which are contained in the bark of the woody stem, range from about 1.2m to 2.1m in length and are extremely strong. Their quality varies depending on the timing of harvesting, and the fibres are graded in terms of their fineness, length, colour, uniformity and strength.

The inner woody stem, the 'shiv' (or 'shive', or 'hurds'), historically has not been used intensively,

Left: Industrial hemp growing in Yorkshire. This plant is ready for harvesting.

but this is changing rapidly in the modern world, with new uses being developed all the time: packaging filler and animal bedding, for example. It is hemp shiv that is used in the production of hempcrete.

The seeds of the hemp plant are used as a food source, and ground to produce oils for a wide range of purposes, including technical and industrial applications. The hemp plant in its whole state can be used as a biofuel, and even the cell fluid of the hemp plant is now used in the manufacture of abrasive fluids.[1]

A history of hemp

The common hemp plant, *Cannabis sativa*, is one of the earliest recorded domestically grown plants, with evidence of its cultivation by humans since Neolithic times. Hemp is found across the world, and has a long history of widespread use for a range of important products: hemp seeds for oils and resin, food, fuel, medicines and cosmetics; hemp fibre for hard-wearing clothing, rope and tough fabrics such as sailcloth (the word canvas derives from cannabis: literally – originally – 'a fabric made from hemp'), and as a pulp from which to make paper.

It is thought that the plant originated in China, and that its cultivation gradually spread westwards through India and into the Middle East, Africa and the Mediterranean, where it formed an essential part of the livelihood and culture of each people who grew it. Surviving writings from the Egyptian, Greek and Roman records show how important the hemp plant was to the lifestyle, trade and expansion of these great civilizations.

The cultivation of hemp in Europe continued throughout modern history, with evidence that its use in Britain, introduced by the Romans, continued thereafter, with the Saxons incorporating it into their medical treatments. Later kings of England promoted the cultivation of hemp, not only for its everyday uses for linen and rope but also for the vital part it played in the military supremacy of Britain as an island nation: Henry VIII passed a law making it compulsory for farmers to grow hemp, such was its importance to the defence of the realm through its use for sailcloth and rigging. Later still, the hemp plant played a not-insignificant part in Napoleon Bonaparte's downfall, since his ill-fated incursion into Russia had as its aim the destruction of Russian hemp plantations. Russia had been supplying the English with hemp, and thus equipping the navy of Napoleon's enemy.[2]

The importance of hemp in British and Irish society throughout the ages is reflected in place names across the land – for example, Hemel Hempstead in the south of England (meaning literally 'place of hemp' or 'hemp pasture'), and Cwm Cywarch in Snowdonia (which translates as 'the steep-walled mountain basin in which hemp is grown'). Street names such as Hemp Mill Walk in Loggerheads in Staffordshire speak for themselves, and Hemp Street in Belfast is at the centre of the area which, until the beginning of the twentieth century, housed the thriving industry around the manufacture of hemp rope and sailcloth for the city's important ship-building trade.

The long tradition of hemp growing and processing in Britain can still be seen in the surviving architecture of the hemp and flax industries. Up until the arrival of cheap imported cotton, sisal and jute, towards the end of the nineteenth century, hemp and flax were still widely used for clothing, other textile products, rope and netting. In fact, these two plants are the *only* fibre crops that are commercially viable in our temperate maritime climate. Many towns in the UK have surviving buildings that were originally a part of

The Ropewalk, Nottingham. Signs of the hemp industry of earlier centuries remain in its surviving architecture across the UK.

these important and widespread industries. The most visible of these are often the 'rope walks' of Victorian, or earlier, times: long buildings where the hemp and flax fibres were stretched out and spun into twine and rope.

For those interested in the history of hemp cultivation in the UK, a good place to start is the quiet coastal towns of Bridport and West Bay in Dorset. The rolling hills of west Dorset and south Somerset, with their well-drained, fertile soil and warm climatic conditions, provide the perfect conditions for cultivating hemp, and the town of Bridport already boasted a thriving hemp- and flax-processing industry by the thirteenth century. The earliest recorded evidence of this is a "payment for a large quantity of sails and cordage in 1211", which was followed by "an order from King John for Bridport rope and cloth to supply the navy in 1213."[3] Although the industry suffered during the sixteenth and

seventeenth centuries, owing to competition from other shipyards and local depopulation as a result of plagues, there was a dramatic revival of the town's hemp and flax production during the period from the late eighteenth to late nineteenth centuries. This, combined with the relatively slow growth and development of the town since the beginning of the twentieth century, means that the history of the British hemp industry is preserved in the buildings of Bridport and West Bay today.

Another of the main uses of the plant throughout history has been in religious ceremonies, and more recently as a recreational drug, due to its relaxing and mildly psychoactive effect. This use of cannabis, or marijuana, as a narcotic eventually led to the growing and possession of the hemp plant being outlawed in most Western countries in the early decades of the twentieth century. The prohibition of cannabis remains in force widely today, in the West at least, with some notable exceptions, such as the Netherlands, which was the first Western country to introduce an official policy of tolerating the possession of small amounts of the drug without prosecution. Other countries, including Spain, Portugal, the Czech Republic and Switzerland, have followed its example in recent years by decriminalizing the possession and use of small amounts of the drug. In November 2012 the US states of Washington and Colorado voted to legalize cannabis, although it is still unclear how this will work in practice, since the drug continues to be illegal under US Federal law, which of course has jurisdiction in both of these states.

Rediscovering hemp

The unfortunate side-effect of the prohibition of the drug cannabis has of course been the blanket banning of cultivation of all forms of the hemp plant, and its consequent unavailability to Western societies for its many non-drug-related uses. The cultivation of hemp in the UK was outlawed in 1928. Since the 1930s, much effort has gone into developing cultivars of the plant which contain very little THC. The success of this endeavour means that, for some decades now, an industrial hemp plant with little or no THC content has been widely available.

The term 'industrial hemp' refers to cultivars of *Cannabis sativa* which have been bred to have a THC content of 0.2 per cent or less. These cultivars have been legal to grow in the UK since 1993, and in Canada since 1995. The THC content in a drug-producing plant would be 10-15 per cent or higher, depending on the strain and method of cultivation. Because the industrial hemp plant looks identical to the drug-producing strains, however, the growing of industrial hemp in the UK requires a special licence from the government.

Since the early 1990s, more and more individuals and organizations throughout the Western world are starting to re-embrace the hemp plant, exploring its potential as a natural and sustainable source of materials with great commercial, industrial, agricultural and environmental potential. The table opposite[4] shows some examples of its many uses.

While the hemp plant is not, as is sometimes claimed, a 'miracle plant' that can solve all of the problems of the Western world, it seems that the resurgence in its use will be invaluable in a future where it is essential that we wean ourselves off our dependence on fossil fuels. At the time of writing, hemp has been grown legally again in the UK for 20 years, and the number of farmers cultivating it is steadily increasing, as local and international demand for the crop expands.

In the United States, the commercial growing of hemp has been made legal under State law in the states of Colorado and Washington, although confusingly it remains illegal under Federal law. In late April 2013, Colorado farmer Ryan Loflin

Hemp cosmetics are now a familiar sight on the high street.

Commercial uses of the hemp plant today

Part of plant		Use
Whole plant		Fuel for boilers Feedstock for biomass pyrolysis
Stem	Shiv (core)	Paper Cardboard Packaging filler Construction (hempcrete, plasters & renders) Animal bedding Mulch Mushroom compost
	Bast fibre (contained within bark)	Quality hard-wearing clothing Bags Shoes Twine Rope Netting Canvas tarpaulins Carpets Geotextiles Fibre-reinforced plastics Construction (quilt insulation, plasters & renders) Brake & clutch linings Caulking
Leaves		Animal bedding Mulch Mushroom compost
Seeds	Whole/ground	Food, including protein-rich flour Birdseed Animal feed (pressed into seed cake)
	Extracted oil	Food (salad oil, margarine & food supplements) Oil paints Solvents Varnishes Chainsaw lubricants Inks Putty Biodiesel
Cell fluid		Abrasive fluids

planted the first major hemp crop in the USA in more than 60 years. This landmark planting, five months after it was made legal in Colorado, is the result of a huge groundswell of opinion in the USA, which has been pushing for the US government to fall into line with the rest of the world in its attitude to the hemp plant.[5]

Hemp farming in the UK today

Hemp grows easily in a range of soils and climates, providing the soil pH is 6.5 or above (neutral-to-alkaline). The hemp plant is not especially nutrient hungry, although for maximum commercial yields a fertilizer is needed. It's also a deep-rooting plant, helping to break up the soil to some depth, which is beneficial for soil health and condition.

The plant is an effective weed suppressant because it grows very quickly and is very 'competitive', winning out over other plants for growing space and light. For this reason it requires no chemical weedkillers, and in fact hemp is sometimes grown specifically to clear land of chemical-resistant weeds. Another useful property of the plant is its pest resistance – there are virtually no pests or diseases that attack hemp – so there is no need to use pesticides or fungicides during its cultivation.

The speed of growth, together with its natural weed and pest resistance, makes hemp a profitable 'break crop' for UK farmers, providing a

useful barrier to pests and diseases in the soil between the sowing of other crops, such as cereals, as part of a crop rotation.

The gradual depletion of nutrients in the UK's soil through the last 60 years or so of ever-more intensive and industrialized farming, and the 'replacement' of these nutrients with increasing applications of chemical fertilizers, has resulted in a very poor soil quality across much of the UK's farming land. Poor soil produces weak plants that are more susceptible to pests and disease, which in turn are targeted with increased levels of chemical pesticides and fungicides. These also indiscriminately kill beneficial insects – both soil-dwellers and airborne pollinators, which are vital for the success of our food crops. Even if the continued use of such chemical fertilizers and pesticides were the best way forward (which it isn't), the fact that many of these products are

Hemp grown in the UK is ready for cutting in August.

derived from fossil fuels means that they are a non-sustainable resource.

The situation nationally at the beginning of the twenty-first century, apart from among those farmers who are returning to smaller-scale, organic farming, is soil across the UK's farms that is almost totally devoid of those insects and other organisms which convert decaying organic matter into the nutrients that plants need to grow. These tiny creatures also provide a key source of nutrients at the bottom of the food chain, and their absence is a significant factor in the huge loss of biodiversity in the UK that has occurred over recent decades. The cultivation of a plant such as industrial hemp, which is not nutrient-greedy, can be grown using only organic fertilizers and without chemical weedkillers and pesticides, and breaks up the deep soil, has the potential to bring huge benefits in terms of soil health, food production and the UK's ecosystem as a whole.

Industrial hemp is sown in the late spring, from early May to early June, at a density of around 30-38kg of seed per hectare. The ideal conditions are a warming soil with plenty of moisture present. Prevailing conditions are more important than exact timing when sowing, since the seedlings are not frost hardy.

Seedlings emerge within five days, and grow rapidly, sometimes at a rate of 30cm a week. The crop is harvested in August, having attained a height of 2-4m. As it is cut down, the hemp plant is also cut into shorter lengths and is left on the ground for up to a month for 'retting': a biological process whereby the hemp straw becomes more workable and the bast fibres begin to separate from the shiv. When the retting process is complete, and the sun has dried the hemp, it is baled and stored under cover. The raw hemp straw is then processed to separate it into saleable parts. The different markets for each part are shown in the table below (see also table on page 19).

Hemp can be grown for the straw alone, or allowed to grow on for a seed crop. It is possible to grow a 'dual' crop, for fibre and seed, although this is much less common because the flowering of the plant does lead to some reduction in the overall quality of the stem, from which the bast and shiv fibres are obtained.

Various models of hemp processing have emerged in the UK in recent years, ranging from the super-high-tech, such as the multi-million-pound processing plant in Suffolk built by Hemp Technology Ltd (formerly Hemcore), to small-scale and (relatively) low-tech solutions such as those employed by K J Voase and Son in Yorkshire, who process all of their own hemp using machinery they have built or adapted themselves, on the farm where it is grown.

Current UK markets for hemp straw

Part of straw	Proportion of crop (approx.)	Market
Shiv (stem core)	60%	Construction industry, horticulture & animal husbandry
Bast fibre (strong fibres in which the stem is wrapped)	30%	Textile industry, construction industry, scientific & technical industries, automotive industry
Fines (small pieces of bast fibre)	7%	Consumer goods (e.g. mattresses)
Dust	3%	Fishing (e.g. ingredient in fishing bait)

CHAPTER TWO

Hemp in construction

Hemp for use in construction forms a relatively small, but growing, proportion of the output from hemp farming in the UK. The main ways in which hemp is used in construction are to make hempcrete and to provide fibres for quilt insulation.

'Hempcrete' is the popular term for a hemp–lime composite building material. It is created by wet-mixing the chopped woody stem of the hemp plant (hemp shiv) with a lime-based binder to create a material that can be cast into moulds. This forms a non-load-bearing, sustainable, 'breathable' (vapour permeable) and insulating material that can be used to form walls, floor slabs, ceilings and roof insulation, in both new build and restoration projects.

Hempcrete was developed in France in the mid-1980s, when people were experimenting to find an appropriate replacement for deteriorated wattle and daub in medieval timber-frame buildings. Across Europe, awareness was growing about the extensive damage that had been done to such buildings in the post-war period through ill-advised repairs using ordinary Portland cement. Using this material to replace the vapour-permeable earth-and-lime mortars and natural cements in historic buildings prevented the buildings' fabric from 'breathing' (see Chapter 4, page 58). This in turn led to the retention of moisture within the fabric, which damaged the timber frames.

A replacement was sought that would not only preserve the vapour-permeable nature of a building's fabric, thereby keeping it in good health, but also provide insulation. It was discovered that the stem of the hemp plant, highly durable and comprised of strong cellulose (capable of going from wet to dry and vice versa almost indefinitely without degrading), was the ideal aggregate to add to lime mortars to achieve this effect. Thanks to the cell structure of the hemp stalk and the matrix structure created by the individual pieces of hemp inside the wall, together with the properties of the lime binder itself, a hempcrete wall has a good ability to absorb and release

Left: Hemp–lime composite – hempcrete – being cast around a structural timber frame.

Hempcrete used as a replacement for wattle and daub.

moisture (see pages 41 and 56). Also, since a great deal of air is trapped inside a hempcrete wall (both within the hemp itself and within the matrix of the hemp shiv in the cast material), it is a surprisingly good insulating material, and the density which the lime binder adds gives the finished material a good amount of thermal mass. Almost as soon as this technique was developed for the repair of historic buildings, people started experimenting with its use in sustainable new build – and found that it was equally suitable for this application.

More detailed information on the way hempcrete works within a building's fabric can be found in Chapters 4 and 7.

Is building with hemp a new phenomenon? It hardly seems likely that human civilizations would have cultivated the plant for millennia for such a wide range of uses *without* using it in their buildings. It is unlikely, however, that physical evidence of any such use in ancient times would survive, since plant-based building materials will of course eventually decay, returning to the soil from whence they came. After all, that is one aspect of the very reason that we are interested in them today: a low-impact building material will allow us to house ourselves 'lightly', without leaving a legacy of adverse effects on the environment behind us.

There is some evidence, however, that building with hemp did not start in the twentieth century and, further, that properly maintained hemp buildings can last for centuries. A historic hemp house in Miasa village, in the Nagano prefecture of Japan, now recognized as a Japanese national heritage site, was built in 1698 and survives in good condition to this day.

At the time of writing, hempcrete has been used in building for around 30 years – since its 'invention' in the 1980s. The use of hempcrete has gradually spread, first across Europe and more recently around the world, and the number of people using it, both in new build and in the repair of older buildings, continues to grow.

In the UK a great many buildings, both commercial and residential, have now been built with hempcrete. A notable upsurge in the commercial use of hempcrete came with the Renewable House Programme, funded by the UK government between 2007 and 2010. Under this scheme, a range of developers received varying levels of public funding to build social housing using natural renewable materials, resulting in the construction of around 200 homes. Of the twelve projects funded, seven used hempcrete as an insulation material.

Since the UK construction industry is notoriously slow to adopt new practices, and has been largely sceptical of the need for (or even the possibility of) using natural materials, State-funded and -driven programmes such as the Renewable House Programme are invaluable in facilitating investigation into issues relating to the large-scale adoption of natural materials within the construction sector. The programme certainly had its challenges, and many of the projects undertaken suffered to some extent from the effects of contractors being given novel materials to work with. However, the overall results were encouraging, and no problems were encountered to suggest that hempcrete, along with other natural materials, would not be suitable for adoption on an increasing scale within the mainstream construction industry.

There are also some other uses of the hemp plant in construction, primarily of the bast fibres in the manufacture of fibre quilt insulation materials, and both shiv and bast fibre as an addition to lime plasters, providing additional strength and some insulation to the plaster. While these materials are not the main focus of this book, they are occasionally referred to throughout, so a brief overview of them is included later in this chapter. First, however, we look at the different ways in which hempcrete can be used.

Hemp shiv for building

Currently in the UK there are no agreed standards for the characteristics of hemp shiv for the construction industry, nor for its production or processing. In France, where the hemp building industry is more established, there are strict guidelines for hemp farmers that govern the quality and colour of hemp shiv to be used in construction, and it is hoped that, in time, similar standards for the nature and quality of the product can be agreed for the UK industry.

The processing of hemp shiv for use as a building aggregate (once all the leaves, seeds and bast fibres have been removed) involves breaking it up into small pieces and removing any remaining fibre

Loose hemp shiv against a new hempcrete wall.

and dust. Hemp shiv for building should be as dry and clean as possible, with a minimum of fines (small pieces of bast fibre) and dust present.

The length of the pieces should be between about 10mm and 25mm, but this is not absolutely critical: successful walls have been made with shiv that contains shorter pieces (especially when spray-applying, see page 29), but it is generally acknowledged that pieces of this length produce a good matrix structure within the wall, which is beneficial for its thermal performance (a material's 'success' in conserving heat and power in a building) and vapour permeability (the degree to which a material allows water vapour through it). Walls are also regularly built with hemp shiv that contains a certain amount of fines, although the proportion must be low, otherwise the fines can soak up too much water and potentially affect the setting of the binder. The absence of dust from the shiv is far more important, since excessive dust can have an even more significant impact on the structural integrity of the wall – in extreme cases leading to collapse (see Chapter 5, page 76). This is because the dust soaks up a very high proportion of the water added at the mixing stage, causing the binder to fail. The only way to avoid this is to compensate by adding a lot more water, but this will significantly extend the drying time of the hempcrete. The presence of excessive amounts of dust in hemp shiv for building is to be avoided at all costs.

Hemp should be stored dry, although it comes in plastic-wrapped bales, so there is a certain tolerance of these being left outside on-site in the short term if suitable storage is not available. The shiv should go in the mixer as dry as possible, to avoid excess water entering the mix, but if some areas of a bale do get wet it will not affect the quality of the finished hempcrete. However, if the shiv has been subject to prolonged exposure to moisture and starts to show signs of rotting (the colour changing to black), all black areas should be scraped out and removed from the bale before adding it to the mix.

There is no need for any treatment of hemp shiv with fire retardants or preservatives as long as it is being used with a lime-based binder for hempcrete. Once cast as hempcrete, the lime in the binder acts to effectively inhibit insect attack and protect from dampness and fire.

Sources of hemp for building

Until recently, the majority of UK hemp farmers supplied their raw hemp under exclusive supply contract to a company called Hemp Technology Ltd (previously 'Hemcore'), which invested in a large hemp-processing plant in Halesworth in Suffolk and supplied hemp products to a range of markets, including the construction industry.

Hemp shiv for building comes from the woody stem of the plant.

Hemp Technology went into administration on 28 October 2013. They appear to be seeking investors, but at the time of writing it seems unlikely that the company will be revived in its current form.

The shiv that Hemp Technology Ltd processed for use in construction was sold on to the market exclusively by the parent company, Lime Technology Ltd, as Tradical® HF and was marketed as part of the Tradical® Hemcrete® system together with Tradical® HB, the hempcrete binder made by industrial lime producer Lhoist, which Lime Technology has the exclusive rights to supply within the UK. To date it has not been possible to purchase Tradical® HF without the equivalent binder, or vice versa, in the UK, and so this has artificially restricted the supply of hemp shiv on to the UK market. However, this may soon change in the light of the current situation at Hemp Technology.

Those wishing to source UK-grown hemp from other suppliers have been restricted to a small number of independent farmers who had not signed exclusive supply contracts with Hemp Technology, and who process their own hemp.

There is nothing to stop builders making links with farmers and sourcing their own shiv for building, but if the farmer is not used to supplying to the construction industry it is important that the essential qualities of hemp shiv for construction are understood by both parties. There needs to be clear agreement in advance about the qualities of the product required (as described earlier) and the cost, including that of transportation to site. Some natural building suppliers supply hemp shiv from independent UK farmers, although this is far from commonplace at the time of writing.

In early 2013 the French company Chanvrière de l'Aube commenced talks with several natural building suppliers in the UK about supplying its hemp shiv through these outlets. Samples shown to the authors were encouraging in terms of the shiv having a very good standard of dryness, a very low dust and fines content, and a consistent colour. However, broad discussions around price suggested that it might be more expensive than shiv currently on the UK market.

It remains to be seen whether importing French hemp shiv into the UK is a commercially viable enterprise. Such imports do not help the development of the (less well-established) hemp-processing sector in the UK, and the use of foreign shiv would make hempcrete less sustainable, since locally grown UK hemp (with, therefore, lower embodied energy – the energy used in sourcing, manufacture and transport of a material) is available for UK builders to use. On the other hand, the presence on the market of French shiv, subject as it is to the more stringent standards imposed by the French authorities, may prove competitive to UK shiv producers in terms of quality, if not price, and may provide the required impetus for the UK industry to develop its own quality standards for construction hemp shiv.

Details of hemp shiv suppliers in the UK can be found in the Resources section at the back of this book.

Cast-in-situ hempcrete

Cast-in-situ hempcrete refers to mixing hempcrete on-site and casting it into moulds constructed from shuttering, or formwork, to form the walls, floor or roof in the exact position that they will remain within a building. The shuttering may be temporary or permanent.

Because hempcrete is a non-load-bearing material, it is always cast around a structural frame, which provides the main load-bearing element of the building. This is usually, but not always, built of timber. This applies whether it is being used in a new build or a restoration context. In new builds the usual method is to construct a simple stud-work frame from softwood, and bury this within the centre of the hempcrete wall, but alterations can be made to the frame to accommodate different design details, both of the wall itself and of internal and external finishes.

Mixing the hemp shiv and binder together with water can be done with a variety of types of mechanical mixer, depending on the quantity needed, the speed at which it is required, the method of application and access to the site.

The freshly mixed hempcrete is either placed (rather than 'poured', since it isn't a liquid consistency), or sprayed into the void created by the shuttering. It is left for a short time to take an initial set (i.e. set hard enough to bear its own weight), after which the shuttering, if it is tempo-rary, is removed and the hempcrete is allowed to dry out gradually over the next few weeks, until it is dry enough for finishes to be applied.

Hand-placing

The hand-placing of cast-in-situ hempcrete refers to the use of manual labour to place the hempcrete into the void created by the shuttering, as well as to ferry it from the mixer to the place where it is needed. The placing process needs to be carried out carefully to ensure both the quality and certain desirable characteristics of the finished material. The manual transport of the hempcrete is done using large tubs or buckets, since it is a relatively lightweight material.

The hempcrete is cast in shuttering, usually temporary, around the structural frame, which is usually placed centrally within the wall. Hand-placing is the 'standard' method of building with hempcrete, although, since it is quite a labour-intensive process, mechanical delivery systems

Mixing hempcrete for hand-placing.

Cast-in-situ hand-placed hempcrete around a soft-wood timber structural frame.

(spray-applying – see below) have been developed. These are particularly suitable for very large-scale commercial applications.

The placing of the hempcrete material by hand allows a high level of control over the quality of the finished product, although there is a need to carefully monitor consistency of workmanship if lots of people are involved in the placing. The low-tech, hands-on nature of hand-placed cast-in-situ hempcrete appeals to self-builders, whether individuals or groups, who have the time to devote their own labour to the build process in order to reduce costs.

The hand-placed cast-in-situ method of applying hempcrete is the main focus of this book.

Spray-applying

The method of spray-applying hempcrete is often used in France and has now arrived in the UK thanks to Myles and Louisa Yallop of The

Limecrete Company. They have been a driving force in the increasing use of sustainable construction methods in the mainstream building industry, and have championed the use first of hand-placed and now of spray-applied hempcrete over recent years. The Limecrete Company's spraying machines, the first of their kind to be used in the UK, have been in operation since early 2012.

As with all things, spraying hempcrete has its advantages and disadvantages. The picture is developing rapidly, with new ways of using the machines, and new equipment to enhance the method, constantly being developed. The spray-applied method for cast-in-situ hempcrete follows broadly the same technique as that for hand-placed hempcrete, but with fully mechanized delivery and placing of the hempcrete, and some minor alterations to the structural frame to accommodate the way the machine works. The hemp shiv is of a finer grade than that used for hand-placed hempcrete (see Chapter 6), as shiv

Spray-applying hemp.

Sprayed hemp is projected against a permanent shuttering board.

Hemp spraying machine.

attached to it. However, since the spraying machine cannot spray around corners, to make the most of this method of application an open, easily accessible frame is required.

For self-builders wanting to carry out the work themselves, sprayed hempcrete is not really an option. The expensive machinery is not something you can hire by the week, and in any case the skill required to operate it can only be developed over time. However, spray-applying has the potential to reduce the number of people needed on-site during the placement of hempcrete, especially on large-scale builds. Against this, of course, you have to factor in the capital outlay, and time and money spent cleaning, transporting, maintaining and repairing the machine.

For builds of up to around 100m³ (a very large house or large community building, for example) the costs of hand-placing and spray-applying are comparable, with any variation usually depending on the frame design and site-specific issues. As you start to move up to larger builds, spray-application really comes into its own.

Pre-cast hempcrete

As an alternative to in-situ casting, hempcrete can be pre-cast into either blocks or framed panels. This usually brings distinct advantages in terms of predictability of the build process, since in most types of pre-cast hempcrete construction the drying of the hempcrete, or at least most of it, is completed off-site, so any uncertainty regarding the time needed for this is eliminated. This is a particular advantage for the schedule of works in large-scale commercial builds, and times when the construction phase must take place during winter.

However, the downside of pre-casting is that it is often a more complicated way of using hempcrete,

containing pieces longer than 20mm was found to block the hose on the spraying machine.

The hemp and binder slurry are mixed together at the nozzle of the sprayer. Due to the force with which it is projected out of the nozzle, spray-applied hempcrete adheres firmly to the surface on to which it is sprayed. Usually this surface is a permanent internal shuttering board, from which the hempcrete is gradually built out to the desired wall thickness in several passes. There is no need for a shuttering board on the other face of the wall – all that is required is the skill to build up a flat wall.

With spray application the structural frame is usually positioned on the inside face of the walls, where the permanent shuttering boards can be

Hempcrete blocks waiting to be laid.

Hempcrete blocks are laid on a thin bed of lime mortar.

with a higher number of processes and materials involved. This in turn means that it may be a less sustainable construction method, although in truth, accurate comparison between cast-in-situ and pre-cast hempcrete is extremely difficult owing to variables such as the scale of the build, design details, the materials used and the distance between farm, processing plant, factory and site. The main pros and cons with pre-cast methods are outlined below.

Blocks

Hempcrete blocks are usually laid by wetting on the surface and bedded using a thin mortar of hydraulic lime and sand. They are coursed in such a way that thermal bridging (also known as cold bridging – a 'cold bridge' is a break in an insulation layer that allows heat to bypass it) between the outer and inner wall surfaces along the mortar joints is minimized. The blocks cut easily with a hand saw, which is useful for fitting them closely around the structural frame, but to improve the speed of construction and to mini-mize wastage the frame should be designed around the block size, or vice versa.

A number of different companies have been producing hempcrete blocks commercially for several years, both in Europe and, to a lesser extent,

Hempcrete blocks in a new-build house under construction. (Note: the timber structure in the fore-ground is wooden scaffolding, not part of the frame!)

Adnams' distribution warehouse was constructed from a mixture of hempcrete blocks and cast-in-situ hempcrete. *Image: Adnams*

in the UK. At first glance blocks might seem to be the obvious way of using hempcrete, especially considering the benefits of off-site drying. However, there is a fundamental inefficiency in casting blocks which are then laid in mortar, because you need a mixer and formwork (shuttering) to make the block in the factory, and then several subsequent processes must take place before it becomes part of the finished building. Casting the wall in situ is more cost-efficient.

The main arguments against blocks are as follows:
- In order to cast blocks of sufficient structural integrity that they will stand up to handling during manufacture, warehousing and transportation, the density of the hempcrete mix must be increased (i.e. the proportion of binder must be higher) compared with cast-in-situ hempcrete. This reduces the insulation performance, although it increases the thermal mass.

- The blocks need to be laid using a bedding mortar which, although only a thin layer, can potentially create cold bridges through the wall.
- The compressive strength of hempcrete is not such that these blocks can be used structurally to support the load of the roof, as concrete blocks would be, so although strong enough to support their own weight, they still need to be laid around a structural frame.

These three factors, combined with the fact that blocks are a more expensive and (potentially) higher-embodied-energy option, mean that cast-in-situ will usually be the preferred method for building with hempcrete.

There have been attempts to cast 'structural' hempcrete blocks, with a higher compressive strength, capable of taking a level of compressive force, but these are not really practical, since the increased density of the hempcrete that is required to provide structural performance means that the insulation value is significantly reduced. Such 'structural' blocks may be most suitable for internal walls, where a higher density is desirable to provide increased thermal mass or better acoustic performance within the building.

Despite their drawbacks, there are many examples of hempcrete blocks being used within the UK to good effect. They are particularly suitable for large-scale builds, for reasons already described, and can be combined with other forms of insulation. A notable example of such a build is the Adnams brewery warehouse in Suffolk, which was built using 100,000 hempcrete blocks combined with 1,000m³ of cast-in-situ hempcrete. Blocks may also be suitable for very small builds or builds where there is limited access, when mixing on-site may be undesirable for cost or logistical reasons.

Panels

At the time of writing, Hemcrete Projects Ltd, part of the Lime Technology group, is the sole producer of pre-cast hempcrete panels in the UK. The panels, which comprise a timber frame-work, a built-in insulation layer and a 'breathable' vapour-control layer, are produced and dried off-site, then supplied and assembled on-site by Hemcrete Projects. In contrast to blocks, there is no need for mortar at the assembly stage, which means no waiting for this to set before finishes can be applied.

Two types of panels are produced, Hembuild® and Hemclad®. The insulation layer in both types is generally hemp-fibre quilt (see page 38), which gives them a lower U-value (better insulation performance) than cast hempcrete for a given wall thickness. This is a development driven by market demand for lower U-values, rather than an indication of natural-fibre quilt's superior overall thermal performance to hempcrete. See

Chapter 7 for a detailed description of the thermal performance of hempcrete.

Hembuild® panels are structural, i.e. the timber-frame elements are designed such that when joined together on-site, the panels form the structure of the building, as well as the insulation and the airtight thermal 'envelope' (the surface that contains the building's internal heated volume). These panels are suitable for one- to three-storey buildings and are probably most suited to large residential buildings, schools and commercial buildings.

Hemclad® panels have an identical make-up except that the timber elements are not designed to be structural. Instead, they are joined together to act as a cladding around a separate supporting structural frame, often made from glulam (glued laminated timber), steel or concrete. These panels are primarily suitable for commercial or industrial

Hemclad® panels are used in this large commercial building.

Panels being made at the factory. *Image: Lime Technology Ltd*

The panels are air-dried at the manufacturing plant, which involves blowing air through them to achieve faster drying, and during the winter this air may be heated. Taken together with the increased use of mechanized processes and transportation, there is some doubt that pre-cast panels can claim the same level of environmental credentials as cast-in-situ hempcrete, and they are arguably less sustainable than blocks because of the additional processed materials they include. However, Lime Technology's literature confirms that the panel systems are carbon negative in terms of their embodied energy.

developments, owing to their cost and scale, but have also been used around oak frames in residential buildings. Both types of panel can be manufactured and supplied in bespoke sizes, and are supplied in a range of U-values, depending on the requirements of the client.

The predictable nature of pre-cast panels, together with their availability in bespoke sizes to fit the building design, is attractive, especially for large-scale commercial builds. While the complexity of panel systems makes them inherently more vulnerable to thermal bridging and poor airtightness compared with cast-in-situ hempcrete – perhaps even more so than with blocks – if the design of the panels is sufficiently

Panels being installed on-site. *Image: Lime Technology Ltd*

robust, these issues can be overcome through detailing and the addition of extra materials into the wall build-up. The published values for thermal bridging and airtightness achieved by Hembuild® and Hemclad® are certainly a demonstration of this. Unfortunately, however, the solution of using expanding tapes between panels to achieve airtightness means the addition of a high-embodied-energy, synthetic material into the wall build-up.

Cast-in-situ versus pre-cast hempcrete

The main advantage of pre-cast panels and blocks is that the hempcrete is already dried before it arrives on-site. This means that plastering can begin almost as soon as the wall construction phase is complete, whereas cast-in-situ hempcrete walls take several weeks to dry sufficiently for finishes to be applied, and this is dependent on local conditions. During the winter months this

drying process makes casting hempcrete in situ a practical impossibility. With cast-in-situ hempcrete, factors such as temperature, exposure, humidity and effective management of drying need to be considered, accounted for and managed within the schedule of the build as a whole. The reliability and consistency of prefabricated panels or blocks are of great benefit in large-scale commercial or industrial applications, where the predictability of how the material will 'behave' once applied on-site allows for reduced costs, and is much easier to fit into a complicated, and often financially critical, schedule of works.

The cast-in-situ method, on the other hand, has the advantage of reduced labour costs because the casting and on-site construction are collapsed into a single process. The process of construction is also simpler, and is achievable with minimal mechanization. Cast-in-situ hempcrete, if it is hand-placed, needs only a mixer and a chain of people to ferry buckets of hempcrete, rather than (in the case of large pre-cast panels) a production

Hempcrete cast up against an outer skin of hempcrete blocks – combining some of the benefits of the pre-cast and cast-in-situ methods.

CHAPTER THREE

An introduction to lime

The main constituent of binders used for hempcrete is building lime: a product made from naturally occurring and plentiful sources of calcium carbonate, it has been used as a building material throughout history.

Over the last few decades there has been a reawakening of interest in the use of lime in traditional buildings. This has led to an increased awareness of the qualities of a range of different building limes, and an understanding of why these are so important – not only in traditional buildings but also in natural building today. There is now a wide range of books on this fascinating and sometimes complex subject, from introductory practical guides to detailed technical manuals, and on the history of building limes (see Bibliography for some examples).

For those wanting to work with lime for the first time, we strongly recommend not only further reading but also an introductory training course (at the very least), since the application of lime mortars and plasters often differs significantly from that of equivalent modern materials. A wide choice of training courses focusing on building with lime is now available across the UK: contact the UK Building Limes Forum (see Resources) for advice and information about courses and workshop events.

The various types of building lime available on the UK market today have a range of useful properties. Certain types of building lime are used when building with hempcrete – in the hempcrete binder itself and usually also in the wet finishes, both internally and externally. The following is an overview of the different types of lime.

Lime in building

Limes have been used in building for many thousands of years. Their importance as a constituent of bedding and pointing mortars, renders and internal plasters, and in the construction of floor slabs, not only underpinned the construction

Left: Natural breathable finishes such as lime plasters have been used for centuries, and are now experiencing a revival.

With skill and patience, lime plasters can be repaired selectively in historic work by patching in new plaster to the old, retaining the areas that were still sound.

It was discovered that certain types of limestone, including some chalks, produced a building lime that was able to set relatively quickly on exposure to water (hydration), instead of slowly on exposure to air (carbonation). These limes also set somewhat stronger, and have a lower level of permeability to water and water vapour than air limes. These properties made these types of lime sought after for use in certain applications where a hard set, a fast set, a set under water or resistance to moisture ingress was required: mainly for significant structural engineering works and for bridges, dams, docks and canals, water tanks and drainage systems. The limes were also extensively used for foundations in damp soils, external renders in damp climates, plasterwork in bathrooms, and as a masonry mortar with certain types of stone (hard stones). Since the 1950s, ordinary Portland cement has been widely used for most of these applications (see page 47).

Because they set through hydration in the presence of water, these limes are known as 'hydraulic limes'. In early times, the hydraulic properties of the various types of stone were usually ascribed to visible characteristics, such as the colour of the stone, but from the eighteenth century onwards there was an explosion of interest and research into the science of building limes, and it emerged that the hydraulic properties were in fact due to the presence of impurities, usually in the form of clay, in certain seams of stone. These impurities, when fired appropriately in a kiln, combine with the lime to form active compounds which in turn react with water to produce the hydraulic set. This set is *in addition* to the setting by carbonation of any remaining 'uncombined' or 'free' limes. These limes are always slaked to a powder because clearly, being hydraulic, they cannot be stored under water like a putty lime.

The impurities present in the stone that create hydraulic compounds may be soluble silica

from seams of limestone that are very pure, containing a high level of calcium and few impurities in the form of clay materials. This pure form of limestone deposit is the most common type found in the geology of the British Isles.

Hydraulic limes

Since air limes set slowly over several weeks or months on exposure to air, they are not suitable for all applications. For thousands of years, masons and engineers have experimented with the use of various types of stone, and found that different types produce building limes with a range of different qualities.

A moderately hydraulic lime mortar is used for pointing in this wall, since the stone is very hard. For softer masonry, an air lime mortar would be appropriate.

(SiO_2), which is the most active, alumina (Al_2O_3), or ferric oxide (Fe_2O_3); the presence of ferric oxide gives a slightly 'buff' colour to those hydraulic limes when they have set.

A range of limes with differing degrees of hydraulicity has always been available, due in part to the differing amounts of impurities present in the original stone, and in part to the temperature at which they are fired. Hydraulic limes are usually classified as 'feebly', 'moderately' or 'eminently' hydraulic, depending on their strength and other characteristics. In feebly hydraulic limes – the type most commonly found in the UK – there is still a lot of 'free' lime in the material, so much of the set happens due to carbonation, with a relatively weak accompanying hydraulic set. In moderately or eminently hydraulic limes, there is far less free lime present, and a stronger hydraulic set is responsible for the majority of the setting process.

Suitable limestone for producing moderately and eminently hydraulic limes occurs less often in the UK geology, and where such seams do exist they have been found to lack the impurities at

consistent levels to produce limes with reliable characteristics. The best moderately and eminently hydraulic limes to be found close to the UK are produced in France.

'Natural hydraulic lime' (NHL), produced from naturally occurring impure chalk and limestone (in contrast to 'formulated hydraulic limes' – see overleaf), is sold in a range of strengths, for example feebly hydraulic (NHL 1, NHL 2), moderately hydraulic (NHL 3.5) and eminently hydraulic (NHL 5). These are sold in specialist builders' merchants in the UK, and more recently NHL 3.5 is also sometimes stocked, or at least available to order, in general builders' merchants. Hydraulic limes are easier to use than air limes for those who have experience of using Portland cement, since their preparation and application is more akin to that of cement.

Natural cements

In limestone in which a very high level of impurities is present, there is sometimes not even enough free lime to enable the lumps of burnt lime to break down on slaking. These types of hydraulic lime are instead ground to a fine powder, to which water is added to form a mortar. They set very quickly and give a very high strength on the addition of water, which has led to their description as 'natural cements'. In fact, natural cements are so quick to set that a retardant chemical is often added at the point of mixing, to give a longer working time.

Such extremely hydraulic limes were used by the Romans in many of their large-scale engineering works, and one particular naturally occurring cement was patented and marketed as 'Roman cement' by James Parker from the end of the eighteenth century. No seams of stone suitable for the production of natural cements exist within the UK; the nearest location in which they are produced is southern France.

One natural cement on the market today, which is now sold as a hempcrete binder, is Vicat's Prompt Natural Cement (see Chapter 6, page 80). Prompt is supplied with a retardant, Tempo (powdered citric acid), which is added on mixing to delay the setting and increase working time.

Other than the difference in production process (grinding rather than slaking) and the extremely fast and hard set, natural cements are largely the same as other hydraulic limes. They retain the important property of vapour permeability, but they are hydrophobic – they repel liquid water.

Prompt Natural Cement is mixed with citric acid to extend its working time.

Pozzolans and formulated limes

Throughout history, wherever hydraulic limes were not available, people have added a range of other substances to air limes that react with the calcium in the presence of water to give the mortar or render a degree of hydraulic set, where this was desirable or necessary. The properties of such 'pozzolans' or 'pozzolanic additives' were probably originally discovered by accident, perhaps through the inclusion of swept-up waste materials on a building site into the mix for a mortar.

These pozzolans all contain a form of very finely divided clay, which combine with some of the free limes present in air limes to produce a

Natural cements were often used in Roman engineering works, such as this Roman watchtower in Nîmes.

hydraulic effect similar to that achieved in naturally occurring hydraulic limes. Examples of pozzolans include certain volcanic ashes (such as pozzolana, found in Pozzuoli, near Naples in Italy, which gave its name to this type of additive), crushed clay brick dust or pulverized fly ash (ash produced by power stations that burn pulverised coal).

The use of pozzolans to create artificial hydraulic limes continues today, especially in the UK, where there has always been a drive to achieve a more hydraulic set than is achievable with locally occurring materials.

Pozzolans can be added to air limes by the builder at the point of mixing or using the mortar or render, and are sold in their raw state from traditional building suppliers for this purpose. In addition, some companies have produced ready-mixed artificial hydraulic limes from dry hydrate of lime (sold as 'hydrated lime') and pozzolans. Such limes are known as 'formulated hydraulic limes' (usually called either FL or HL – to distinguish them from the 'natural' NHL), and are usually sold through specialist building material suppliers.

Portland cement

The increased interest from the 1750s onwards in the properties of different building limes led to many artificial limes being synthesized and patented in an effort to create mortars with particular desirable characteristics.

The most significant development was the patenting of Portland cement, by Joseph Aspdin, a Leeds bricklayer, in 1824. Portland cement is formed by the burning of specific quantities of limestone and clay-containing materials at very high temperatures. Since it has many of the same setting properties as natural cements, it took a

long time for Portland cement to replace lime in the construction industry, owing to its initial high cost and the fact that limes had always been available locally. But its use really took off from the early twentieth century, and by the post-war period of the 1950s and 1960s, ordinary Portland cement had all but replaced hydraulic lime for mortars, renders and engineering applications in UK construction.

The main advantage of Portland cement over most building limes was its hard and predictable set, giving mortars with greater structural performance and strength in tension compared with the softer-setting lime. Also, cement's shorter curing time meant that builders could go on working into the colder months; masonry construction had traditionally been a seasonal activity, dependent on the minimum temperatures required while lime mortars and renders slowly re-carbonated. Building in the colder months meant an increase in productivity and profits, and if the weather became colder still, the proportion of cement in the mix could be increased from 4:1 to 3:1 (sand:cement) to decrease curing times still further. Of course these advantages were already available to builders through the use of natural cements, but since these tend to have a natural buff colour on drying, they were not ideal for one of their major uses: the imitation of stone in, for example, decorative cornices or statues. It was the desire to create a cement with a similar grey colour to that of many stones that was the driving force behind the first innovation in artificial cements.

Significantly, in contrast to those made from lime, mortars and renders made from Portland cement have very little permeability to moisture vapour. This characteristic was initially welcomed, and cement enthusiastically used to apply a 'waterproof' barrier, both in engineering works and to houses and other buildings as an external render.

Cement causes damage to natural materials because it is not breathable enough. Here, bricks originally laid in lime mortar have been repointed in Portland cement, which is harder than the brick. Any water that penetrates into the face of the brick, instead of wicking away into the softer mortar, now stays there – and when it freezes it expands, blowing the face of the brick.

The merits of the ubiquitous use of Portland cement in the new construction of the post-war period is itself questionable with hindsight, but its use for the repair of older buildings – originally built with lime mortars and renders – was disastrous. The application of cement stopped these buildings from working properly, allowing moisture to become trapped in the fabric of the building and leading to the rapid deterioration of the natural materials used in their construction.

Since the 1980s there has been a growing understanding of the damage done to traditional and historic buildings through the use of cement and, as we saw in the last chapter, there has been a huge resurgence in the use of lime mortars and renders in older buildings, especially in historically significant buildings, where the importance of lime in preserving these structures in good repair is now widely recognized.

From the point of view of those interested in sustainable building, limes are also preferable to cement for a number of other reasons:
- The burning temperatures needed for the production of limes are lower than for cement (approximately 900°C compared with around 1,400°C), so their embodied energy is lower in that respect.
- Those limes that set by carbonation *reabsorb* carbon dioxide during the setting process, thus taking *out* of the atmosphere at least some of the CO_2 that was given off during the manufacturing process.
- Limes can help to ensure a healthy indoor air quality in buildings, reducing the need for powered air-conditioning systems.

Gypsum

At about the same time as cement took over from lime in mortars and renders, the UK saw the use of lime in plastering almost completely replaced by gypsum.

The appeal of internal plasters made from gypsum (calcium sulphate, $CaSO_4$) instead of lime was primarily their ease of workability and fine finish. Gypsum plasters had been available for centuries, but were previously too expensive to be used anywhere other than in the grandest houses. Industrial quarrying and mass production methods in the early twentieth century, including

Lime and gypsum plasters side by side. The finish of the cream-coloured lime plaster around the window appears more textured and warm than the smooth gypsum on the surrounding walls and ceiling.

the ready availability of cheap pre-cast gypsum plasterboards, suddenly made gypsum plasters cost-effective in the post-war housing boom.

Like lime, gypsum plasters are hygroscopic, with a good ability to absorb moisture vapour. However, unlike lime, which allows moisture to pass in and out of it easily without significant damage to the structure of the plaster or mortar (such finishes are designed to be reapplied eventually), gypsum *holds on* to any moisture it absorbs. Moisture is locked within the structure of the plaster until it reaches saturation level, at which point the plaster's structure collapses and it falls off the wall. For this reason, gypsum plasters, usually used together with non-vapour-

permeable paints, were not really a successful replacement for lime plasters in older buildings. Like cement, they contributed to the building not working as it was intended to (see Chapter 4, page 58).

Limes: summary

While debate has always raged (and continues to do so) in both the heritage and sustainable building industries about which type of lime is the 'most authentic' or 'best', such arguments are largely meaningless. A quick glance at the historical use of lime teaches us that building limes were always used, as they are today, in a range of forms to suit a range of requirements. The different characteristics of the different types of lime are each appropriate to different applications. The variety of limes available is perhaps best viewed as a kind of 'toolkit' of binders for use in mortars, renders and plasters, combined with a wide range of different materials. Anyone who claims to have a preferred form of lime is either missing the point or only ever does one kind of work.

In considering building limes, the key issue to understand, other than the techniques involved in applying lime plasters and mortars (which are varied and sometimes complex, and beyond the scope of this book to discuss in detail), is that of vapour permeability and the way in which this helps to keep the building in good health. See page 58 for more on this subject.

An important rule of thumb when considering the construction of a wall is to have harder, less permeable substances towards the centre of the wall, and softer materials, with greater permeability, towards the outside. This favours the transportation of moisture out of the fabric of the wall. It would therefore be advisable, taking a stone wall as an example, to use a moderately

CHAPTER FOUR

Key concepts in sustainable building

Before we go on to look in detail at building with hempcrete, it is worth briefly examining some key concepts and ideas that underlie the use of this construction material. A good grasp of these ideas will help to inform sensible choices during the design and construction phases of the build process.

Although most people who choose to use hempcrete today generally have a good understanding of these concepts, at present, in the UK at least, these vital factors are largely ignored within the legislative and regulatory framework relating to the construction industry. This means that there is also limited awareness of these ideas in wider society, which is an impediment to the wider use of truly sustainable construction methods.

Zero-carbon buildings?

We live in a world increasingly aware of the need to minimize fossil-fuel use and limit the emissions of carbon dioxide into the Earth's atmosphere. The construction industry in the UK is responsible for an estimated 64 per cent of all UK carbon emissions.[1] This enormous carbon bill comes both from the energy consumed in the construction of the built environment and from that used by the buildings' occupants during the lifetime of those buildings.

Clearly there is an urgent need to move to methods of building, and construction materials, that are responsible for less CO_2 emissions in their manufacture and use. In addition, to reduce the energy consumed during a building's lifetime (most significantly in heating and, in hot

Left: Hempcrete house in a new development in Oxfordshire, clad in local stone to reflect the local vernacular.

insulation combined with a good amount of thermal mass. In comparison, cob or rammed earth has high thermal mass but virtually no insulation value, and straw bale has good insulation value but virtually no thermal mass. Hempcrete is unique in providing both good insulation *and* good thermal mass.

The slow release of the heat stored in a hempcrete building's thermal mass, when the heat source is turned off, means that there is less fluctuation between extremes of temperature inside the building. The thermal mass in the walls *buffers* changes in internal temperature, meaning that the indoor temperature in a hempcrete building is slower to respond to changes in the outside temperature (increasing cold *or* heat) than in a building with less thermal mass.

This aspect of its thermal performance, combined with its good insulation value, makes hempcrete very efficient at regulating and maintaining internal temperatures. The building's heating and/or cooling system therefore uses less fuel than it would in an environment in which heat is lost quickly and the indoor air frequently has to be brought back up to a comfortable temperature from a low starting point.

For information on research into hempcrete's thermal performance in a building, see Chapter 7.

Thermal bridging

As described in Chapter 2, cast-in-situ hempcrete is wet-mixed and then cast as a single insulating mass, in a shape dictated by the construction of shuttering. It thereby forms a continuous flow of material around the walls – and sometimes the floors, ceiling or roof – of the building.

This feature of hempcrete as a building material, when cast in situ, allows for the easy elimination of thermal bridging, or 'cold bridging'. This often

Cast-in-situ hempcrete forms a continuous mass, reducing thermal bridging.

occurs at points around the structure of a building, where a material with higher thermal conductivity forms a connecting passage across the insulating layer, from the outside air to the inside. Such 'bridges' allow the escape of heat from the indoor environment through the 'tunnel' of the less insulating material, and often result in localized cold spots where condensation can occur, promoting mould growth.

With hempcrete cast as a continuous mass, it is quite straightforward to design out cold bridging in a hempcrete building, compared with a building whose walls are constructed from a build-up of different materials. The most challenging area in which to avoid cold bridges is the insulation of the plinth at the foot of the hempcrete wall.

Airtightness

Airtightness is an important consideration in the construction of buildings to a modern standard. The increasing emphasis on the conservation of energy has led to a focus on the amount of energy wasted in 'leaky' buildings, through heat that escapes through holes that allow the passage of air. In conventional construction, certain materials, for example membranes, are used for the specific purpose of adding an airtight perimeter to the interior heated space of the building. Due to the complexity and multi-material nature of modern construction, however, this airtight perimeter is vulnerable where sections of this membrane overlap and are joined (usually by a tape), and at junctions with other materials.

The fact that hempcrete forms a continuous mass within the building's walls makes it comparatively easy to ensure airtightness. Although hempcrete is naturally porous, with an open surface, it achieves good levels of airtightness when plaster and/or render finishes are applied.

Due to its good thermal mass, hempcrete allows the use of natural ventilation of the interior space (see below right). This avoids the risks associated with high levels of airtightness in conventional buildings built with non-porous synthetic materials, and brings benefits in terms of air quality.

For more information on airtightness in a hempcrete building, see Chapter 22.

Indoor air quality

As we saw earlier in this chapter, synthetic insulation materials can cause health problems for a building's occupants, as a result of chemicals they contain 'off-gassing' into the indoor atmosphere. This is not an issue with hempcrete, which contains no toxic materials.

We have also seen that hempcrete is valuable in helping to manage humidity levels within the building because of its hygroscopic nature. This too is important for indoor air quality, because maintaining internal levels of humidity has important health consequences. If the humidity level drops too low, there is an increased risk of allergies and asthma, and if it is too high, the risk of the growth of moulds, fungi and mites rises. There is an increased risk of bacterial or viral infections, especially of the respiratory tract, if the humidity rises above *or* drops below the ideal range of 40-60 per cent relative humidity. 'Relative humidity' describes the amount of water held in a body of air, at a given temperature, expressed as a percentage of the maximum amount of water that body of air would hold if completely saturated.

The thermal mass of hempcrete also plays a part in achieving a good indoor air quality, because of its implications for ventilation practices (see box on page 311). In short, the indoor air quality in hempcrete buildings is excellent. In summary, this is achieved by:

- the hempcrete, together with lime or clay plasters, acting in a hygroscopic way – absorbing and releasing moisture and keeping the humidity of the indoor environment within levels optimal for human health
- there being no toxic chemicals or VOCs present in hempcrete to off-gas into the air
- the thermal mass of hempcrete enabling the storage of heat in the building's fabric, which is only released very slowly when the temperature of the nearby air cools, thus allowing some use of natural ventilation (i.e. opening windows) to maintain indoor air quality, without all the heat escaping.

These factors combine to ensure that a hempcrete building provides a natural, healthy and comfortable living environment with little or no recourse to mechanical ventilation systems.

CHAPTER FIVE

Getting the basics right

The main reason for constructing walls out of hempcrete is to equip the building with a very high standard of insulation without resorting to the use of synthetic insulation materials, which have a high embodied energy and are produced from non-renewable materials – usually petrochemical derivatives or synthetic fibres.

Hempcrete is a relatively new construction material, about which the general public is only just becoming aware. Perhaps because of this, we at Hemp-LimeConstruct frequently get enquiries, even from people seriously considering using the material in their building, from which it is clear that the basic principles of building with hempcrete are not fully understood. In the case of clients who are employing someone else to do the construction work, clearly there is a limit to the level of detail they actually need to know about the process. There are, however, certain concepts that they will need to understand in order to maintain their hempcrete building in good order.

For those considering building with hempcrete themselves, however, it is obviously important that they know what they are doing, and that the basic principles are understood at the start, so that mistakes are avoided later on. This chapter, therefore, provides an overview of the key principles of hempcrete building. Some of these themes have been introduced in previous chapters, and we return to them all throughout the book, but gathering them together here should serve to highlight their importance. At the end of the chapter is a brief overview of problems that can occur with hempcrete as a result of lack of understanding of the material and the construction process.

The nature of the material

Hempcrete is essentially a modern version of very old natural composite construction materials, such as wattle and daub or cob. While it is relatively low-tech in its composition and

Left: A freshly cast hempcrete wall.

application, the quality of workmanship through-out the construction phase can make a huge difference to the finished material.

The material is formed by wet-mixing hemp shiv (the chopped woody stem of the industrial hemp plant) with a lime-based binder. The cast-in-situ method involves casting it in 'moulds' made from temporary shuttering, which is taken down as the hempcrete begins to dry. Hempcrete is also available as pre-cast blocks or panels, but cast-in-situ hempcrete is the focus of this book.

Since hempcrete is not strong enough to be load-bearing, it is always cast around (or, in the case of pre-cast, built around) a structural frame, which supports the weight of the upper floors and roof and provides stability to the hempcrete. When subjected to excessive force in testing, hempcrete tends to 'fail' by bending, rather than breaking, and this flexibility of the cast material can be a very useful characteristic, for example enabling a hempcrete building to tolerate small ground movements over time.

Hempcrete has very good thermal performance, partly owing to the insulation provided by the air trapped within the cast material, and partly owing to the relatively high thermal mass of the finished material, which means it can store heat. For more details, see Chapter 4, page 62.

The lime present throughout the cast material helps to protect the various elements within the finished wall from rotting when exposed to moisture, and also inhibits insect attack. Hemp shiv is not, in any case, an attractive food source for insects.

Hempcrete provides a means of creating walls, ceilings and floors that are insulating, yet are made from sustainable materials (hemp is a renewable, locally grown plant, while lime is derived from abundant naturally occurring material and has a lower embodied energy than cement). It is particularly relevant to those wanting to build in a low-impact, sustainable way, and for upgrading the thermal perform-ance of historic buildings in a way that works

Hempcrete is always cast around a structural frame.

Making use of hempcrete's excellent thermal performance: a walk-in hempcrete fridge.
Image: Margot Voase

harmoniously with the original fabric of the building.

Although the oldest hempcrete building is only a few decades old, there is no evidence to suggest that properly maintained hempcrete will not last for the lifetime of the average building, if not significantly longer. However, even in the event of a failure, since such buildings always contain a load-bearing structural frame, the hempcrete could be removed and re-cast without catastrophic effects on the building as a whole.

Because the lime binder has a lower embodied energy than cement, and because the hemp plant takes up carbon dioxide while growing, which is then locked up in the building, hempcrete acts as a 'carbon sink'. In other words, even accounting for the embodied energy in its production,

transport and construction, a hempcrete wall represents negative carbon emissions: it is responsible for a net reduction rather than a net increase in atmospheric CO_2. This process is what is referred to as carbon sequestration. Timber products can also claim to be sequesters of carbon; however, hemp is superior to wood in this respect, since it absorbs CO_2 much more quickly, creating a very hard woody stem (2-4m in height) in only 4-5 months. The time taken for the production of timber is much longer, even for fast-growing trees.

A range of figures on the carbon sequestration of hempcrete is available, although further research is needed into this area to bring clarity. For example, one 2003 study suggests that a total of 325kg of CO_2 is stored in 1 tonne of dried hemp.[1] Lime Technology cite the following net carbon sequestration figures for their Tradical® shiv and binder system: sprayed hempcrete sequesters 110kg CO_2 per m^3 of hempcrete construction, and shuttered and hand-placed hempcrete sequesters 165kg CO_2 per m^3 of hempcrete, depending on the level of compaction during construction.[2]

Both the lime in the binder and the structure of the hemp plant itself contribute to hempcrete being a 'breathable', or vapour-permeable, material. This means that any water that gets into the wall is able to get out again, rather than being held inside it, where over time it is likely to cause damage to the building's structure and also reduce the thermal performance of the wall. Furthermore, both the lime and the hemp are hygroscopic, which means that the surfaces of the wall are able to absorb moisture from the air during humid times, and release it again when the air dries out. This has the great advantage of helping to maintain the quality of the indoor air, and is thereby beneficial to the health of the building's occupants. For more on this see Chapter 4, pages 58 and 65.

The breathability of the finished material must be maintained over time through the use of vapour-permeable finishes such lime plasters and render, and vapour-permeable paints or limewash, so it is important that the occupants of the building are aware of this.

While people have been building with hempcrete for decades, and thousands of hempcrete buildings now exist globally, the industry is still learning about its potential, developing new binder products and working towards agreement on best practice about construction methods. To some extent, every new hempcrete building at the current time is 'experimental' in that it adds to the growing body of evidence of how this material performs.

The build

The process of building with hempcrete is, of course, the subject of this entire book, but the fundamental principles are set out here. For a more detailed overview of the construction process, see Chapter 11.

It is important to understand the way in which adherence to the correct application technique (or otherwise) can affect the success of the build. Building with cast-in-situ hempcrete is essentially a three-stage process: the erection of a structural frame, the wet-mixing and casting of hempcrete around that frame, and the application of finishes when the hempcrete is sufficiently dry.

The main causes of problems are to do with the amount of water introduced at the mixing stage. In simple terms, there is a 'battle' for water in the mix between the dry hemp shiv and the powdered lime-based binder. The binder needs water to achieve a set, so there is a risk, if too little water is added, of a failure of the hempcrete caused by the binder not setting properly. On the other hand, too much water in the mix can cause problems by significantly extending the drying time of the wall. This can result in delays with the application of finishes, which in turn can slow down the progress of the build and end up costing time and money. These problems are discussed in more detail later in this chapter – see page 73.

A key skill in hempcrete building is knowing exactly how much water is needed, and maintaining a consistent high level of accuracy when measuring out ratios of hemp:binder:water as each mix is prepared. In addition, the builder needs to understand this process well enough to confidently and safely adjust the amount of water slightly in response to weather conditions.

Once the wet hempcrete is mixed, it is placed into temporary shuttering, which creates the shape of the finished wall like a mould. As each 'lift' is cast and takes its initial set, the shuttering is removed and moved up the wall to create the next lift, until the top of the wall is reached. The distinction between 'initial set' and 'drying

Freshly mixed hempcrete should not be too wet. The method of testing for the correct water content is discussed in Chapter 15.

time' is an important one. Various types of hempcrete binder are available, but each specifies a short period of time (usually overnight) after which the binder has set sufficiently that the cast hempcrete can support its own weight. The shuttering is then removed to allow the drying to start. The 'drying time' refers to the time taken for any excess water introduced during mixing to dry out of the wall. It normally takes several weeks until the wall is dry enough for finishes to be applied (see Chapter 16, page 215), so it is important not to increase this time through the inadvertent addition of surplus water.

The phrase 'drying time' does not mean that the wall dries out completely. In common with all natural materials, hempcrete always has some water within it. It dries until it reaches a natural equilibrium or 'resting moisture content', which fluctuates slightly depending on environmental conditions, because of the vapour-permeable and hygroscopic properties of the wall.

The other important aspect of application technique is the placing of the hempcrete itself. (Hempcrete is 'placed' rather than 'poured'.) The wet mix needs to be placed evenly and consistently within the void created by the shuttering. In particular, it is important to avoid over-compaction of the cast material. Over-compacting reduces the insulation performance of the finished material and also increases costs, as more hempcrete is needed to fill the void. In addition, the consolidation of the external faces of the cast hempcrete needs to be judged carefully, as this can also have implications for drying time, as well as the cost of finishes (less well-consolidated surfaces may allow faster drying of the wall, but require a greater thickness of plaster or render).

The above factors mean that, perhaps more so than with other natural building materials, variations in the technique or skill of the builder can have a significant effect on the progress of

the build, and on the performance of the finished product.

The issues described above are discussed in more detail in Chapters 15 and 16.

The workforce

With the importance of technique in mind, it is worth turning our attention briefly to the people who will be doing the building.

Building with hempcrete is, at least on a building of any reasonable size, quite a labour-intensive process. This means that it can often be quite a fun and sociable experience, and since with a little training the process of casting the hempcrete itself is quite easy, pretty much anyone can get involved with that stage. The low-tech nature of the material and the opportunity to reduce costs

Mixing and placing hempcrete is a low-tech process.

by providing their own free labour makes hempcrete especially attractive to self-builders.

Skills and experience

A range of different skills are required at the different stages of the building process, but, on the whole, once the key principles are understood, the actual process of construction should be well within the reach of most people. Building the foundations, wall plinth and structural frame (see Chapters 12 and 13) requires skills that most general builders, carpenters and competent DIYers already possess. The typical studwork structural frame for cast-in-situ hempcrete is either completely encased within the hempcrete or at least hidden from view in the finished wall, so it is not necessary for this to be a beautiful feature with a high standard of joinery. As described earlier, the process of mixing and casting the hempcrete itself, while not exactly complex, requires a certain level of knowledge and skill to ensure mistakes are not made. However, it is also at this stage that best use can be made of inexperienced volunteer labour (see opposite).

When it comes to applying finishes, the skills needed will depend on the type of finish chosen. Finishes range from lime plasters and lime renders to cladding in a variety of materials, for example timber, brickwork or stonework. The application of lime plasters and renders is a distinct skill, different from working with gypsum plasters and cement render, and it should not be assumed that just because someone is skilled at the one, they will be able to do the other. That said, although lime plastering is a specialist trade, depending on the level of finish desired, many self-builders may also be tempted to have a go at doing this part of the work themselves. For more on this, see Chapter 18.

In developing the skills of the team, training courses may be a first practical step. Courses in building with hempcrete are available (see Resources), though still few in number. Courses in lime plastering are now readily available all around the UK, and attending one is recommended for those with no experience who intend to do their own plastering.

Training in lime plastering can be a worthwhile investment.

Whether or not it is possible to gain training in building with hempcrete, some first-hand practical experience, for example volunteering on someone else's build, is certainly advisable for those wishing to use this material for the first time. Such opportunities can be invaluable, because hempcrete is so 'different' from other construction materials, so the chance to see how it works and to experience the practicalities of a build is the best form of preparation. If this is not possible, we would recommend that you at least practise by building a small structure such as a garden shed or workshop, to experiment and perfect the technique before beginning the main build. In any case, before casting for the first time, it is important to construct a small section of frame and cast a test wall, so the technique can be reviewed and mistakes corrected before starting on the actual walls.

The different roles on a hempcrete build are explored in more detail in Chapter 19.

A note on volunteer labour

One of the many appeals of hempcrete is that costs can be kept to a minimum if volunteer labour is used. Many self-builders sensibly take the approach of working on the job themselves and getting friends involved during the build. However, if volunteers are to be used effectively, it is important to think carefully in advance about what tasks they will be asked to do. The issue can be complicated due to volunteers often coming and going through the course of the build rather than staying for the whole time, but it is important that some thought is given to basic training of everyone.

Given the importance of consistently high standards of workmanship for the success of the hempcrete material, as outlined in this chapter, it should be clear why there needs to be a balance between free labour and experienced personnel

on-site. As well as providing basic training and an explanation of the key principles on a hempcrete build, it is important to maintain an appropriate level of supervision over inexperienced workers, whether paid or volunteer. At Hemp-LimeConstruct we have found that one experienced worker to three volunteers is an effective ratio; above this, it is hard to ensure that consistent standards are maintained.

Problems with hempcrete?

Over the past few years, as more hempcrete buildings have been built in the UK, there have been a number of reported problems with the material, which have caused some people in the sustainable building world to take the view that hempcrete 'doesn't work' or 'never dries out'.

These assumptions are obviously incorrect, and all the more unfortunate since in most cases these 'hempcrete problems' are entirely avoidable and are not the fault of the material itself but rather of the contractor who used it. Furthermore, the vast majority of these problems – which usually centre on slow drying of the cast material – are a short-term inconvenience rather than a catastrophic disaster. It would be a shame for hempcrete to get a bad name in the construction industry owing to problems that would not have occurred if there had been proper understanding of the material, and we hope that books such as this one will help to prevent this.

It is true, however, that when people with little experience of lime and/or natural building materials first try to build with hempcrete, problems can occur. Rather than ignore examples of bad press for hempcrete arising from these issues, we feel it would be useful to explain why they can happen, and to reinforce the message that a well-trained workforce which understands the materials is the only sure way to avoid contractor error.

Any problems that do occur are likely to be caused by one of the following factors.

Contractor error

This is the most common cause of problems, especially the problem of slow drying. Most of the perceived problems with hempcrete are in fact the result of builders being unfamiliar with lime, and with natural, plant-based materials, and in particular with the procedure for mixing and placing hempcrete. Issues occur because insufficient thought has been given to the time of year, weather conditions, drying times and management of drying once the material has been cast.

Well-publicized examples (including on TV programmes) have included novice hempcrete builders casting new-build houses on large-scale commercial projects during the winter months, using a slow-setting binder and applying finishes too early.

Incorrect amount of water

In particular, it is vital to understand the importance of adding exactly the right amount of water when mixing, as too little can result in the binder not setting properly, and too much can lead to excessive delays in the hempcrete drying. For more on this, see Chapter 15, page 192.

Applying finishes too early

Where this occurs, it is usually due to pressures of the build schedule, combined with inexperience of the contractor. The waiting time for the wall to be dry enough to plaster will be unnecessarily extended if too much water is added at the mixing stage, so contractor error at that early stage can have knock-on effects at the end of the process.

All finishes for hempcrete should be vapour permeable, so the cast hempcrete will continue to dry out through the finish. However, by applying finishes we reduce the amount of ventilation to the surface of the wall, especially with plaster/render finishes, which also reduce the wall's surface area. These two effects together can considerably extend the amount of time required for the wall to dry out, meaning that it takes longer to dry to moisture levels that are optimal for thermal performance, and occasionally that the internal environment is too damp for the building to be occupied.

In addition, if plasters and renders are applied too early then staining of these finishes can occur, as liquid moisture evaporates from the surface and leaves tannins, which have been carried through from the hemp shiv. This is a purely cosmetic problem, and the stains can be painted or limewashed over when the wall has dried sufficiently,

An extreme case of staining on plaster as a result of finishes being applied too early.

but it is unsightly and can incur extra costs (in terms of reapplying paint). In more extreme cases (see photo opposite), it is also a clear indication that there were serious contractor errors at the construction phase, and/or that the finishes have been applied when there is still a very high level of moisture in the wall.

For information on drying times and how to gauge when a hempcrete wall is ready for the application of finishes, see Chapter 16, page 215.

Faulty materials

This is a much rarer cause of problems, and one that could of course occur with any building material, but in practice it should hardly never happen, as long as materials are bought from a reputable source.

Faulty binder

A faulty binder might either not take a sufficiently strong initial set (so the wall is unable to support its own weight, and slumps downwards) or not cure properly over time, leaving a wall that is structurally unsound. Both of these problems are virtually unheard of in hempcrete construction except in the case of materials that have not been properly tested, such as the two described as follows.

This is the reason that we don't recommend making your own hempcrete binder according to one of the various recipes that have been made available online or in books (see Chapter 6). If you really have a desire to experiment with making your own binder, then treat it as just that – an experiment. *Never* use untested materials to construct a real building.

Worryingly, however, we at Hemp-LimeConstruct have recently seen two examples of complete material failure: one concerning a material sold

for use as a hempcrete binder by a UK company and one concerning a complete ready-mixed 'hemp-limecrete' sold by an Irish company to a UK self-builder. In neither case did the material manufacturer-supplier accept any liability for the failure of the material, and in fact the Irish company refused to do a site visit or to accept recorded correspondence on the subject sent by the UK client.

In the first example, the desire (for reasons of sustainability) to develop a hempcrete binder that contained UK lime and no Portland cement led to a new product being sold to UK hemp-crete builders for use in buildings as a kind of 'in-practice test', with the consent of clients. Unfortunately, the binder material in question was completely inappropriate for hempcrete and failed even to hold its own weight, despite being built up to height quite slowly. After a week or two, gaps appeared at the top of the wall where the material had slumped down, and after a further two weeks a bright multicoloured mould appeared all over the surface (see photos below and overleaf)

Hempcrete wall made with a newly developed (and insufficiently tested) binder, two months after casting.

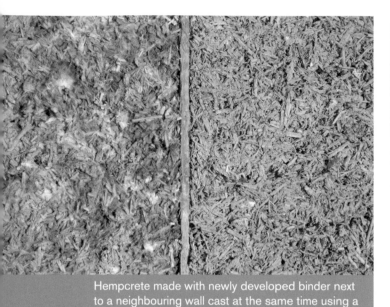

Hempcrete made with newly developed binder next to a neighbouring wall cast at the same time using a proper hempcrete binder.

'Hemp-limecrete' one year after casting.

and in the voids left where the material had slumped. The wall had to be removed at the expense of the builder and client, since the manufacturer refused to accept any liability for the material.

In the second example, a full year after application the 'hemp-limecrete' bore no similarity to any hempcrete we have ever seen, despite being recommended for exactly the same uses in the manufacturer-supplier's technical sheet. The material was clay-coloured and visually resembled an earth wall containing pieces of whole hemp plant, and a year after building it was still registering 100-per-cent moisture levels and had black mould on the surface, with mushrooms growing out of it (see photo above right). The material was so soft that a finger could be pushed into the surface with light pressure. The end result was that the UK self-builder had to pay to have all the material removed from the wall and the frame altered and treated with fungicide before contracting us to re-cast the walls using proper hempcrete materials.

Unfortunately, the current state of the hempcrete market in the UK, with its rapid growth in demand for hempcrete materials in the absence of any widely agreed industry standards, provides the ideal environment for irresponsible or unscrupulous people to market materials that have not been subject to rigorous standards of production and testing. Until such UK industry standards are developed, builders wishing to use hempcrete in the UK would be well advised to stick to the three binders described in this book (see Chapter 6). If you are considering the use of an alternative hempcrete material, always ask to visit buildings that have been built using it, and ask to speak to previous satisfied users of the product.

Faulty hemp

Poor-quality hemp in the mix can also lead to problems, mainly if an excessive amount of fines or dust is present in the hemp shiv. This is because both fines and especially dust can absorb excessive amounts of water, which can deprive the binder of the water it needs to take its initial set (see Chapter 15, page 192). This potentially causes a weak bonding of the hempcrete material, leading to a wall that slumps or even collapses entirely. The failed 'hemp-limecrete' material described on the previous page is advertised on the manufacturer-supplier's website as containing 'coarsely milled whole hemp including long fibre', which in our opinion is completely inappropriate as an aggregate for hempcrete.

In 2012 Hemp-LimeConstruct, along with others in the industry, were supplied with a batch of hemp that contained an excessive amount of dust. The result of this was that after ten days of placing hempcrete in a new-build property, the wall we had cast on the first day suddenly collapsed. This meant that the entire 40m³ of hempcrete already cast had to be removed from the frame and re-cast using good-quality hemp shiv. The cost of replacement materials was borne by the supplier, but the removal and disposal of the old hempcrete and the reinstatement works were at Hemp-Lime-Construct's cost. Thankfully, our clients, while understandably very upset, didn't lose faith, either in hempcrete or in us!

Problems: summary

Our intention in including this section about hempcrete 'problems' is not to scare people away from using it; nor do we want to contribute to any idea that hempcrete is very difficult to use. In fact, once a few simple techniques and concepts are mastered, the construction process is relatively simple. Furthermore, the development of national (and international) standards for the production, testing and application of hempcrete materials will, it is hoped, put a stop to cases of faulty materials – and materials purporting to be hempcrete which in fact bear little resemblance to it – being sold on the UK market.

In fact, if proof were needed, the thousands of hempcrete homes and commercial buildings built in the UK, Europe and around the world over the last three decades stand testament to the fact that hempcrete is a viable construction material that for its success simply requires a level of understanding and competence from those wishing to build with it.

The catastrophic effects of excessive dust in hemp shiv.

CHAPTER SIX

Variations on the hemp–lime mix

In this chapter we look in more detail at the hemp–lime mix itself, outlining the variables in both the hemp shiv and the binder that can affect the characteristics of the finished hempcrete, and describing how the mix is adjusted for different applications.

Variations in the hemp shiv

In general, there should not be much variation found in the type of hemp shiv supplied for building, which should meet the requirements as set out in Chapter 2. That is, the pieces should be about 10-25mm long, and may contain a certain amount of fines, but as far as possible should not contain any dust (see page 26). The only variation commonly found is that a finer-grade shiv (with shorter pieces of shiv) is commonly used to make hempcrete for spray application, as the longer particles in the standard shiv can sometimes lead to blocking of the spraying hose.

When placing by hand, coarse-grade shiv should always be used, since the matrix these create when they are set in the wall provides an inherently stronger and more open (breathable) structure. The shorter pieces of hemp in spraying shiv do not create the same matrix structure, but the additional adhesion given them by the forced projection of the spray-applied hemp compensates for any loss of structural strength.

Variations in the binder

There are two stages to the setting of a hempcrete wall. The first is the initial set, which has to happen before the shuttering can be taken down. Because the shuttering needs to be removed as early as possible so that the hempcrete can start drying out, the initial set needs to be strong enough that the freshly cast hempcrete can support itself while the second stage takes place. This happens slowly, over a number of weeks, as the hempcrete sets fully and any excess water dries out of the wall.

Left: A lime, or lime-based, binder is used with hemp shiv to form hempcrete.

At the risk of stating the obvious, the binder is the part of the hempcrete that is responsible for both stages of the set. Although the binder for hempcrete is lime-based, choosing a binder is not simply a case of choosing your preferred lime, since there are several functions required of the binder, and most limes will not provide all of these by themselves.

A hempcrete binder needs to:
- provide the initial set to the hempcrete with enough strength to support the weight of the drying hempcrete wall (it needs a strong hydraulic set)
- allow the water to continue to dry out of the hemp shiv after the initial set (it needs vapour permeability)
- provide a full set to the hempcrete over time (it needs long-term structural strength).

Most types of building lime, including the feeble and moderately hydraulic limes, have a slow initial set (over several days) and take a long time to reach full strength. Even once the initial set has taken, most limes do not, by themselves, set sufficiently strongly at an early enough stage to support the weight of a hempcrete wall while it dries.

If a hempcrete wall is built with a lime that does not achieve a quick enough, or strong enough, initial set, the consequences can be severe. When the shuttering is taken down, the lower parts of the wall are vulnerable to compacting and slumping outwards at the face of the wall, or – in the worst-case scenario – collapsing completely (see Chapter 5, page 77).

To avoid this, it is essential that the binder contains something that ensures a fast, strong set within 24 hours so that the shuttering can safely be removed. Not only must a successful hempcrete binder achieve this strong initial hydraulic set, but also it must remain vapour permeable during and after setting, so that moisture can leave the hemp shiv and the wall as a whole. This double function is achieved in one of two ways by binder products currently on the market.

First, there are two proprietary binders specifically formulated for hempcrete, which are available in the UK at the time of writing. Although their exact ingredients remain commercially confidential, they are widely thought to be a mixture of air lime (hydrated lime) – the main ingredient – together with a certain amount of Portland cement, believed to be around 20-30 per cent, depending on the manufacturer, and possibly a small amount of other pozzolanic ingredients (see Chapter 3 for more information about lime, cements and other additives). The proportion of cement included is enough to give the required initial strong set, without being enough to inhibit the overall vapour permeability of the cast hempcrete. These proprietary binders are Tradical® HB, made by Lhoist, and Batichanvre®, made by St Astier. They were both developed in France, although Tradical® HB is now made under licence in the UK.

The second solution to achieving the required initial set is to use an extremely hydraulic naturally occurring lime, such as a natural cement. One such natural cement, Prompt, made by Vicat, while originally sold for other purposes, has also been tested as, and is now marketed as, a binder for hempcrete. Natural cements set very quickly and very hard, but retain the vapour permeability that Portland cement lacks. Prompt Natural Cement is made from a specific type of limestone, using the same process as that used to produce hydraulic lime, and is made in France, where this type of stone is found.

There is a further characteristic of hempcrete binders containing a high proportion of air lime that is worth noting here. Since air limes set very

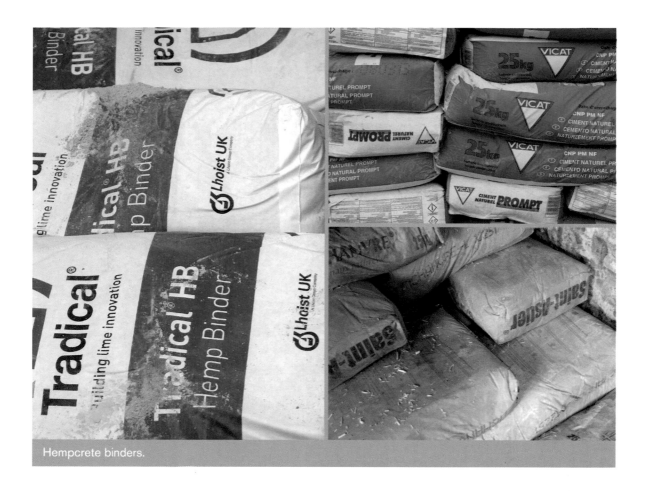

Hempcrete binders.

slowly, through a process of carbonation with the air, and since the water introduced during mixing has to gradually evaporate out of the wall, it is usual for hempcrete made with such a binder to shrink slightly – often by 2-3mm at the edges – as it dries. This is not a problem for the cast hempcrete material per se, but it does have implications for airtightness in certain situations (see Chapter 21, page 296).

The main characteristics of the three binders are set out in the table overleaf.[1] At the time of writing, these three products are the only tested and reliable binders available for hempcrete building in the UK, but there are signs that the market is opening up. We understand that Vicat is currently in the process of formulating a

hempcrete-specific binder based on Prompt Natural Cement, and we know of at least two other companies that are also in the process of developing a hempcrete-specific binder.

Some of the early literature on building with hemp included recipes for mixing your own hempcrete binder. This is not something we have ever done in a commercial context, and nor would we recommend it, for the following reasons:

- It is much harder, when mixing dry ingredients yourself on a small scale, to ensure a consistent mixture comparable to that achieved in a binder that has been blended using purpose-built machinery.
- Although potentially cheaper in terms of materials, it adds extra complication, and

CHAPTER SEVEN

Performance of hempcrete in a building

The performance of hempcrete as a building material is the subject of a growing body of research, both in the UK and abroad. However, because the material is still relatively new, more research is needed for a full understanding of the way it behaves and which factors influence its performance. Each new hempcrete building provides an opportunity for us to further develop and standardize good practice in hempcrete construction, and to gather more evidence about how the material performs.

Plant-based products in general are set to be a growth sector supporting a wide range of technologies, as the world makes the transition into a low-carbon economy. Within this context, the viability of hempcrete along with other bio-aggregate building materials is being studied in a number of research centres; notable examples are the University of Bath in the UK and the Université de Rennes in France.

Much of the research to date has been undertaken, or funded, by the manufacturers of proprietary products and has focused on proving the performance of their own products in relation to regulatory regimes for building products, which vary slightly from country to country. The comparison of data and research findings is complicated by the use of different materials, and different techniques for their preparation and application across different studies, as well as by differences in the particular focus and legislative context of the research. Furthermore, there has been reluctance from some quarters to make proprietary research available to a wider audience,

Left: Hempcrete acts as a continuous, breathable, solid mass of insulation.

CHAPTER EIGHT

Tools and equipment

The equipment needed when building with hempcrete varies slightly from project to project according to a range of factors, including the specific application and construction details, the type of binder used and the scale of the job (as well as how much of the rest of the build you may be undertaking). However, the core tools remain the same, and a list of our standard toolkit is provided below.

Most of the hand and power tools required are commonplace and will already be owned by most builders and keen DIYers, and the bigger power tools and plant can always be hired for the duration of a project. But there are bound to be a few new purchases to be made when building with hempcrete for the first time, and a little thought at the outset may result in significant savings, especially when it comes to power tools. That said, there are some tools on which it is worth spending extra money.

Quality versus cost

The relative cost of electric power tools has fallen dramatically in recent years, especially with the proliferation of cheap generic models, which are churned out in the Far East and stamped with the 'own-brand' label of the DIY store they end up in. The low price is a reflection of the quality of the manufacturing, and often of the quality of work achievable with them. Sadly, the cost of these inferior products has now fallen so much that we often hear people talking of how it is cheaper, for example, to buy four cheap planers to do a job, throwing each one away when it dies, than to buy a model from a good-quality brand.

Of course, even if we move up a level from throwaway tools, there is always a balance to be struck between cost and quality when buying any equipment, whether hand or power tools. With experience comes an awareness of which tools (for any given task) need to be of high

Left: A small mixer is useful for knocking up plaster.

CHAPTER NINE

Health and safety

Health and safety often gets a bad press nowadays, especially in the context of the overzealous attitude many organizations have to it in today's increasingly litigious culture. At best, health and safety at work is not the most thrilling topic, and this chapter is unlikely to be the first that readers turn to when they pick up this book. However, we would emphasize that there are some very important health and safety factors to be kept in mind when building with hempcrete.

Of course, working on any building site is potentially dangerous, with many hazards that have the potential to (and do) cause accidental injuries and deaths. Common risks include trips and falls, falls from height, accidents with power tools and machinery, exposure to hazardous substances, crushing injuries, head injuries and electrocution, to name but a few. This has resulted in the development, over the last 40 years, of a range of legislation, working practices and procedures aimed at improving safety by

reducing the risk of accidents as far as possible, and minimizing their impact on individuals when they do happen.

Plenty has been written about health and safety on building sites, and we do not intend to reproduce it here. Suffice to say that it is the responsibility of everyone on-site to ensure that, in line with Health and Safety Executive guidance, safe working practices are followed at all times, the correct personal protective equipment (PPE) is used, materials are safely stored and tools and machinery are operated safely by trained personnel.

The main health and safety issues associated with hempcrete are to do with the nature of the materials; specifically the lime binder. Because these issues may not be immediately obvious to those who have not worked with building lime before, it is important to ensure that everyone is aware of them before starting work on a hempcrete build. Reading the advice in this chapter may give you the impression that working with

Left: Gloves for mixing or placing hempcrete.

CHAPTER TEN

Planning the build

Before moving on to look at the construction process in detail, in this chapter we briefly discuss some aspects of building with hempcrete that are helpful to keep in mind during the planning and building design stages.

For a more detailed examination of the factors involved when designing hempcrete buildings, see Chapter 21.

Inspiration

Anyone in the fortunate position of being able to design their own building is likely to draw their inspiration from a range of sources. In theory, the only limit is what your imagination can do within the constraints of the materials you are using, although in practice local and national planning policies, building control and consideration towards neighbours will also restrict what is possible in any given situation. Whether the new building is to be a home, workplace or community building, or a smaller living space such as an extension or garden building, the first step in the design process is the spark of an idea about why the building should take a particular form or look a certain way.

With the explosion of interest in self-building that the UK has experienced over recent decades, those designing their own building are hardly short of information about other people's projects from which to draw inspiration for their own plans. The choice of hempcrete as a building material should not limit you in the type of building you design, although the decision to use it would suggest that you are aiming to design an energy-efficient, low-impact building.

While hempcrete is used to create a wide range of types and styles of building, it is our hope that most people seeking inspiration for designing with hempcrete will take as their starting point the natural materials involved, and so look for inspiration to other buildings in their local area that have been constructed with sustainable,

Left: A newly renovated seventeenth-century cottage, with external hempcrete solid-wall insulation.

Focus on self-build 1: Agan Chy

Inheriting a field with a dilapidated concrete barn on it gave Bob and Tally Moores an opportunity to fulfil what was, for Bob at least, a long-held dream of building his own house. Planning permission had been refused on the barn once, but after some detective work Bob and Tally found an old map which convinced the council that, contrary to previous opinion, the barn sat within the village boundary, and permission was eventually granted for demolition of the barn and construction of a house.

Bob is a carpenter, used to timber framing, so his house was always going to be based around a green oak frame, and having worked for a supplier of traditional and environmentally friendly building products, he was familiar with lime and other natural materials. He was doing an MSc in Sustainable Architecture at the Centre for Alternative Technology, from which he says he learnt a lot, but couldn't understand why everyone was talking about easy-build bolt-together timber-frame houses with lightweight insulation, which needed to be sealed up tight to keep the heat in and then ventilated using mechanical systems to maintain indoor air quality. Bob says, "Up to the point when we built the house, I had lived mainly in vernacular buildings, built from local, natural materials that had stood the test of time. I wanted the same feeling of permanence from the house I was going to build myself . . . to know that, as well as being a high-thermal-performance eco-house, it would stand a chance of being there for centuries to come, and I didn't think I would get that from a lightweight insulated timber-frame house." The more Bob thought about it, the more it seemed to him that thermal mass was the key to passively storing heat, whether created by heating systems or from the sun, and slowly releasing this energy to maintain a constant comfortable temperature inside.

The final design included a green oak structural frame, with a softwood studwork frame built off this to take the hempcrete. The principles of passive solar design were followed: highly efficient glazing on the south-facing elevation and a minimum of windows on the north side, together with a good overhang so that the windows are shaded in summer but allow direct sunlight in during the winter, when the sun is lower in the sky. All the external walls are 300mm hempcrete, with additional thermal mass provided by a deeper-than-usual concrete floor slab and slate floor covering in the open-plan living area. The roof is insulated using wood-fibre insulation panels. Being in an exposed location close to the Atlantic coast in north Cornwall, Bob and Tally have sensibly used a larch rain-screen cladding over the hempcrete on the exposed walls (most of the house); on the south side, they used a breathable render. Bob's motto for the build was 'Low-tech – high performance', and this comes out in the solidity and strength of the materials that surround us as we stand in the kitchen: oak posts and beams, solid black slate flooring, black slate windowsills, lime plasters and thick hempcrete walls. The house, on a scorching July day, feels reassuringly cool and comfortable, despite the fact that we are sitting next to the large south-facing windows and the external doors are open, allowing a direct connection with the heat outside.

In the winter, heating is provided by a wood burner in the living room, the flue of which

1. Agan Chy.

2. 'Solid and built to last', yet light and airy.

3. Larch cladding was used as a finish for the hempcrete.

4. Natural materials complement each other at Agan Chy: natural slate sills and roof covering, and lime renders or larch cladding over hempcrete.

5. Warm in winter and cool in summer, despite the large windows, due to hempcrete's unique thermal performance – insulation *and* thermal mass – together with the passive solar design.

passes through the master bedroom on its way to the roof, allowing the passive transfer of heat into the bedroom. There is a solar domestic hot water system, and an air-source heat pump to supply underfloor heating, but Bob and Tally find that they hardly ever need to use it. Even in very cold winters, the heat from the wood burner alone is more than enough to maintain a comfortable temperature throughout the house "at least 90 per cent of the time – and to be honest a lot of the time we probably have it on for the atmosphere rather than heating. The only time we turn the heating on is when we've got guests in the spare bedroom and we feel we should heat up the 'north wing' for them," says Bob.

Bob designed the house with architect Roderick James, and the hempcrete details were worked out by Bob together with Chris Brookman from Back to Earth, who supplied the hempcrete. The build took eight months of full-time work, and another eight months part time. A friend, Ollie, worked with Bob on the construction of the frame, and then Bob and Tally finished the rest of the build by themselves. Bob says he found the hempcrete easy: "It's great to work with – putting up the shuttering was the hardest bit." He enjoyed the low-tech nature of hempcrete as a material, and found it refreshing being a pioneer doing something that not many others had attempted at the time, and so being able to "make it up as I went along" – a pleasant change from conventional construction.

At the time, Bob worked out what it would have cost to build the house out of brick-and-block, and "it was pretty much the same – and that includes the 'cost' of my own labour, but if you think of the advantages from the thermal performance of hempcrete there's no comparison". Bob and Tally found that using hempcrete brought no disadvantages in terms of getting a mortgage, and there were no issues with planning either. For building control they used an Approved Inspector, and

would recommend this to others: "Because he was directly employed by us, he took the time to listen and understand the material and what we were doing, and instead of being suspicious of something out of the ordinary he was interested and really got behind the project."

Reactions from friends and family have been interesting: "In the early stages, lots of people were curious, if not openly amused," Bob says, "but once they have experienced what it feels like to be in the house, they all get it. Everyone is blown away by it." Bob's musician friends particularly enjoy the acoustics created by the solid hempcrete. "Not being a musician, it's an aspect of the material I'd never really thought about," says Bob, "but they rave about it." Anyone wishing they could experience the feel of a hempcrete house for themselves can do so, if they fancy a summer holiday in Cornwall, as Bob and Tally rent their house out for eight weeks of the year.

Following completion of the build, Bob was surprised to get a personal call from the energy assessor who had visited their house. "He said that he never rang people up, but he was calling to advise me to get an airtightness test done, because ours was one of the most thermally efficient houses he had ever assessed, and with favourable airtightness results it might put our rating up from A+ to something even higher – the very highest rating, which only a very few buildings in Europe ever achieve." Bob has not in fact had an airtightness test carried out, as "I didn't feel it was worth spending the money, and because we wanted to use passive ventilation we've got trickle vents. I'm not really interested in super-airtightness if it means using powered mechanical heat-recovery ventilation systems; anyway, we're happy with how it performs – it's warm in winter and cool in summer, and it achieves my criterion of feeling like a solid house – something that's built to last!"

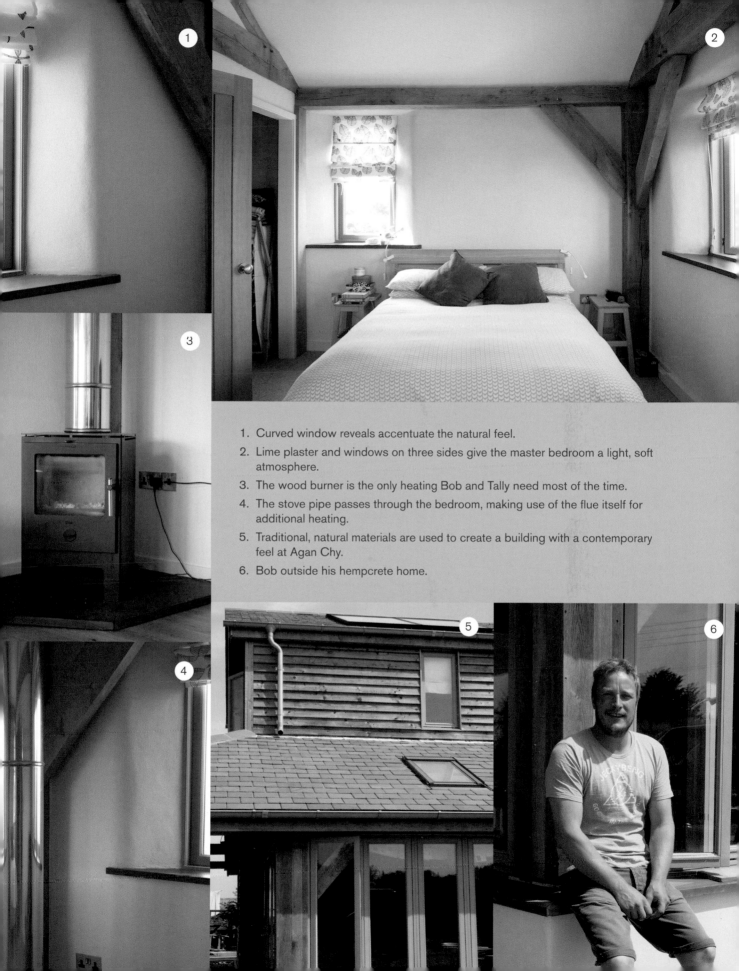

1. Curved window reveals accentuate the natural feel.
2. Lime plaster and windows on three sides give the master bedroom a light, soft atmosphere.
3. The wood burner is the only heating Bob and Tally need most of the time.
4. The stove pipe passes through the bedroom, making use of the flue itself for additional heating.
5. Traditional, natural materials are used to create a building with a contemporary feel at Agan Chy.
6. Bob outside his hempcrete home.

PART TWO

Hempcrete construction

CHAPTER ELEVEN
The hempcrete wall: an overview 131

CHAPTER TWELVE
Foundations and plinth 147

CHAPTER THIRTEEN
The structural frame 155

CHAPTER FOURTEEN
Shuttering 169

CHAPTER FIFTEEN
Mixing hempcrete 191

CHAPTER SIXTEEN
Placing hempcrete 207

CHAPTER SEVENTEEN
Floors, ceilings and roof insulation 221

CHAPTER EIGHTEEN
Finishes for hempcrete 231

CHAPTER NINETEEN
Practicalities on a hempcrete build 253

CHAPTER TWENTY
Restoration and retrofit 267

CHAPTER ELEVEN

The hempcrete wall: an overview

There are several ways of using hempcrete within the structure of a building, but its 'standard' use, which is probably the most familiar to those who have seen or worked with the material, is the construction of solid (monolithic) walls to form the thermal envelope.

This chapter describes the principles of construction and the typical construction method for a cast-in-situ hempcrete wall. There are other ways of using the material in walling, depending on the specific application and the specifications of the architect or building designer, but the basic principles are the same.

Later chapters in this part of the book provide an expanded discussion of the methods used and skills involved at each stage of wall construction. Other than the information given in this chapter on services, all topics relating to the wall build-up are covered more thoroughly in the rest of Part Two. The purpose of this chapter is to give a broad overview of the entire walling system.

Hempcrete wall construction principles

A hempcrete wall is built with natural materials and provides a very high level of thermal and acoustic insulation and is 'breathable', so provides moisture buffering – passive regulation of humidity, which is beneficial both for human health, because it improves indoor air quality, and for the fabric of the building. Different levels of insulation can be achieved by casting different thicknesses of hempcrete, but a standard thickness in new build would typically be 300mm or 350mm. The thermal performance achieved varies according to the exact materials used.

As a monolithic cast-in-situ walling system, hempcrete effectively minimizes the chance of thermal bridging, by forming a continuous sheet of insulation material all around the building. The wall build-up, in its simplest form, uses only two or three different materials (lime, hemp, timber), which adhere closely to one another,

Left: Hempcrete internal walls in an Elizabethan timber-frame building.

Glulam roof structure in a large hempcrete building.

and it contains no cavity, thus minimizing the risk of interstitial condensation.

More information about the benefits of using natural insulation materials and the thermal, moisture management and other properties of hempcrete can be found in Chapters 4 and 7.

Hempcrete walls are cast around a structural timber frame, as they are not strong enough to be load-bearing. Although they have some strength in compression, this is not enough in itself to support the weight of the roof and upper floors. The set hempcrete, though, does provide a good strength in tension, which provides

racking strength to the timber frame, meaning that the use of timber can be minimized in a well-designed frame.

It is perfectly possible to construct large buildings, including those with structural frames of glulam, steel or concrete to accommodate several storeys or cross large spans, using cast hempcrete. However, these are more likely to involve complicated frame designs, and in the case of very large buildings it is likely to be quicker and more cost-effective to use prefabricated hempcrete panels or blocks or sprayed cast-in-situ hempcrete than to place the hempcrete by hand.

For buildings of up to two or three storeys, hand-placed cast-in-situ hempcrete provides a natural, sustainable and low-tech walling system which, though relatively labour-intensive, is easy to construct and is comparable in cost to conventional masonry. Big savings can be achieved by self-builders and community groups who can provide their own labour.

The fact that the hemp aggregate is a natural plant material has implications for the design of the building, for which 'a good pair of boots and a good hat' is generally recommended. In other words, the building should have a good-sized masonry plinth to lift the bottom of the hempcrete above the ground, to protect it from standing

A large eaves overhang is not always necessary on hempcrete buildings: at Callowlands in Watford, a hidden gutter and a specialist hydrophobic lime-based render allow the overhang to be eliminated.

water or being constantly wetted by splashback from rain hitting the ground, and a good roof overhang to throw rain away from the wall. In practice, however, many hempcrete buildings have been built with alternative, compensatory eaves details without causing problems. The degree of roof overhang should be considered carefully alongside the choice of external finish for the walls.

These factors, together with the requirement to keep all finishes breathable, the extended drying time, and the need to give some thought at an early stage to fixing into the finished walls, are probably the elements that will be the most unfamiliar to designers and builders of conventional buildings. Other aspects of hempcrete construction, such as the timber-stud framing, the shuttering, the mixing of a binder with an aggregate, and the plaster or cladding finishes, will be familiar to most people in the construction industry.

Internal walls

Hempcrete is often specified only for the external walls of a house, because of its insulative properties, but there is no reason why it cannot be used for internal dividing walls as well. Hempcrete internal walls create a continuation of the 'feel' of the building into the interior space, and provide a good thermal and acoustic barrier between rooms. Interior hempcrete walls can be cast at a much-reduced thickness, and they make it easy to avoid the use of non-natural or high-embodied-energy materials such as gypsum plasterboard and synthetic insulation to construct stud walls.

Using hempcrete for internal walls also allows the continuation of the use of natural finishes such as lime or clay plasters throughout the interior space, although in many cases architects apparently do not consider this, or see no benefit

A good-sized plinth keeps the hempcrete wall in this new-build house well above ground level.

ground. Hempcrete's low density allows some reduction of the amount of concrete used in some foundations, thereby reducing the embodied energy of the build (most commonly, where the ground calls for a raft foundation instead of the standard trench – see Chapter 12, pages 148 and 150-1). Free-draining foundations can be very useful in hempcrete buildings, to reduce the potential for moisture being held against the bottom of the hempcrete wall.

At the base of the walls, built off the foundations, is a plinth, usually of masonry construction, which serves as the 'good pair of boots' for the building. The plinth has a damp-proof course (DPC) running along the top of it.

For more detail about the foundations and plinth see Chapter 12.

The structural frame

On top of the plinth and the DPC, the structural frame is constructed. The usual method is to build a simple studwork frame consisting of a floor plate, studs and a wall plate of untreated softwood timber. The cross-sectional dimensions of the wood required vary depending on the size of the building, and of course for buildings of several storeys or which include wide ceiling spans, a steel or glulam frame may be necessary. All fixings must be alkali-resistant (stainless steel or, in the case of nails, hot-dip galvanized), as they need to resist corrosion by the lime in the hempcrete (see Chapter 21, page 299).

Further details on timber-frame design are provided in Chapter 13.

The frame is usually encased centrally within the cast hempcrete wall, but can be positioned flush with the surface of the hempcrete, either internally or externally, to make fixing into the wall easier. For example, a wall that is to be clad with timber or masonry externally would benefit from the frame being flush with the external face of the wall. Likewise, in a room where a lot of heavy-duty fixings will need to be made into the wall (e.g. for hanging kitchen wall cupboards), then it is normal to have the frame flush with the internal face of the wall. Positioning the frame flush with either surface of the wall necessitates some additions to the frame to ensure that the other two-thirds of the wall are properly supported.

A note on fixing into hempcrete walls

Hempcrete, when properly applied, should not be tamped hard into the shuttering, but instead spread evenly around in the void with a gloved hand (see Chapter 16, page 211). This should result in a finished wall of low-density ('light and fluffy') hempcrete, which has a high insulation value because of the air trapped in the wall. A low-density hempcrete also dries more quickly.

Despite the 'open' matrix structure created by the particles of binder-coated hemp within the wall, it sets very hard and strong when fully dry, and is perfectly strong enough to fix into for light-duty fixings, for example for pictures or small mirrors, using wall plugs as you would with a conventional masonry wall.

For more heavy-duty fixings, such as wall-mounted radiators, guttering downpipes and wall-hung cupboards, a hempcrete wall may not have sufficient density to support the weight hanging from it. There are two solutions: the first is to use wooden rails cast flush with, or just behind, the surface of the wall. Including a horizontal wooden rail in the frame design at, say, 900mm and 1800mm enables a strong fix anywhere in the room at dado and picture-rail heights. Clearly this needs to be integrated into the design at the wall build-up stage; if this is not done, and strong fixings are required once the hempcrete has set, the solution is to use wooden wedges hammered into the dried wall to take a fixing. These are set so that the wide end sits flush with the wall surface and screws can be fixed straight into the wedge, and have the advantage that they can be set in at the exact place where the fixing is required. You may have seen this method used in Victorian houses as a way of providing a fixing for skirting boards. It is particularly effective with hempcrete, since the set hempcrete has elasticity and 'bends' under excessive pressure rather than breaking. This means that the hempcrete around the wedge gets pushed outwards, and becomes denser in the immediate area surrounding the wedge, thus holding it more firmly.

As an alternative to fixings, remember that, since hempcrete is a cast material, it is relatively simple to create attractive and functional recessed alcoves and shelving by making a mould to use as part of the shuttering and casting the hempcrete around it. With this in mind, take the time to plan your use of each room carefully before you start work.

Where the timber frame is buried centrally in the wall, it is well protected, by the vapour-permeable nature of the hempcrete, from moisture being held against it. Furthermore, the lime in the binder not only protects against rot but also acts as a barrier to insect infestation and as an anti-fungal agent. It is therefore acceptable to use untreated softwood for parts of the frame that are encased in the wall (see page 134).

Where the frame is flush with the external face to assist with tying cladding to the wall, this part of the frame needs extra protection from water ingress. For details, see Chapters 13 and 18. Where the frame is flush with the internal face, we have always used untreated wood, but this will be a decision for you and for your building control inspector. An internally exposed frame is obviously less vulnerable to damp ingress, but

Hammering a wooden wedge into a hempcrete wall.

Downpipe fixed into hempcrete with wooden wedges.

depending on the types of insect prevalent in your area, you might still have to consider treated wood or an appropriate hardwood. The timber may be covered with 10mm or more of lime plaster, which itself gives a decent level of protection from insects for as long as it remains intact.

The frame design is a part of the process that is well worth spending time on, in order to gain the maximum benefit from possible variations of design around the building. Frame design needs to take all of the above factors into account, and the designer also needs to think about the temporary shuttering system at the frame design stage, so that it is clear how the two will interact. This aspect is often overlooked, perhaps because designers rarely have direct experience of building with hempcrete.

Shuttering

Shuttering for hempcrete walls is either temporary, constructed on-site from a lightweight reusable timber board (11mm OSB is usually the preferred choice), or permanent, constructed from any breathable carrier board that will take a lime or clay plaster over it, and which will not laminate away from the surface of the hempcrete behind it (typically a wood wool board).

Temporary shuttering is fixed to the frame using very long heavy-duty screws, and is held at an exact distance off the frame by spacers (usually made from plastic waste pipe cut to length). Permanent shuttering needs a frame detail which includes studs positioned on one side of the wall, to which the shuttering board is fixed.

After the placed hempcrete has taken its initial set (usually overnight, but the time varies according to the binder used), temporary shuttering is struck, and moved along or up the wall to the next place it is needed. In this way the total shuttering board required on-site is restricted to a little more than the amount needed to shutter up what you can fill in a day and a half (see Chapter 14, page 171).

Care must be taken to ensure that shuttering is true, plumb and square, so that the face of the finished wall is straight and vertical.

For more detail about shuttering, see Chapter 14.

Openings for doors and windows

Openings for doors and windows are created as part of the structural frame, or by an additional

The framework for a window is left flush with the surface inside the reveal to provide a fix for the window frame.

The shuttering boards used to form the window reveals are fixed to the framework for the window.

Angled window reveals are attractive within a deep hempcrete wall and maximize the amount of light.

CHAPTER TWELVE

Foundations and plinth

The foundations for a hempcrete building are similar to those required by any other, with some minor differences, which we explore in this chapter. The plinth, which sits directly on top of the foundation, is really a part of the wall itself, but it is significantly different in nature from the hempcrete part of the wall.

In many respects the plinth has more in common, both conceptually and in its materials, with the foundation than with the hempcrete above it. For this reason these two elements are discussed together in this chapter, which explains their roles in providing a sound basis for the hempcrete structure above them.

Foundations

The foundations for your building will usually be specified by an architect and/or structural engineer or geotechnical engineer, and built to that specification by you or your contractor. The building control inspector will check the ground conditions on-site once the excavation stage is complete, and advise if further depth is required.

The foundation exists for the purpose of transferring the load (weight) of the building safely down to stable ground. Foundations for a hempcrete building follow broadly the same principles as for any other building, although the low density of hempcrete as a walling material does allow for other possibilities in the type of foundation that can be used. This may translate into a lower embodied energy, since a foundation usually demands high fossil-fuel usage, as it involves casting several tonnes of concrete.

Concrete foundations

The most common type of foundation is called a 'strip foundation'. This means digging a trench along the footprint of the load-bearing walls until solid ground (e.g. firm clay or rock) is reached, and then filling this with concrete to create a solid footing.

Left: The plinth for this hempcrete building is constructed from natural stone and lime mortar.

CHAPTER THIRTEEN

The structural frame

Hempcrete is not a load-bearing material, so it is always used in conjunction with a structural frame to take the load of the roof and upper floors safely down to the foundation. The hempcrete is typically cast around the frame, and provides racking strength to it. For buildings of up to three storeys, the frame will generally be made from softwood timber. Hempcrete is perfect for casting around timber frames as it has a certain level of natural elasticity and is able to cope with the slight movements that are normal in timber-frame structures.

Various types of structural frame can be used with hempcrete, and the frame is usually designed by, or at least its structural members specified by, a structural engineer and is subject to close scrutiny from building control. In larger buildings, where the frame is likely to be glulam (glued laminated timber), steel or concrete (see overleaf), a softwood sub-frame is still required to support the hempcrete. You will need to provide your engineer with a set of drawings containing at least floor plans, elevations and your intended wall section, showing where ideally you would like the structural frame to be positioned. Your engineer will also need to know what type of roof covering you are going to have and the position of any heavy mechanical or electrical equipment intended for the roof (e.g. solar panels). He or she will also need details of the density of the hempcrete you are using, and will need to understand the implications of casting hempcrete around a timber frame in terms of reducing horizontal frame members. From this information the engineer will be able to specify the size of structural members within the frame and provide a set of calculations to satisfy the appropriate Building Regulations requirement.

Your engineer may not have come across hempcrete before, so if you want to simplify the frame, to reduce costs and make placing the hempcrete easier, you may have to inform him or her about its characteristics. The frame design has the potential to make the process of placing the hempcrete very easy or very difficult.

Left: A softwood timber frame for a new-build hempcrete extension.

Large buildings

A very large hempcrete building usually has a substantial structural frame, especially where it has either a very heavy roof or is supporting four or more storeys. Large frames can either sit within the hempcrete wall or be set away from the wall inside the building. It is beyond the scope of this book to discuss the design of massive structural frames in detail, and those who are constructing such buildings will have engineers to design a frame with the suitable structural capabilities. For engineers seeking to adapt such frames to be suitable for hempcrete, the principles outlined below can be followed.

There are three likely options for the frame:
- **Steel.** The main consideration with steel is the risk of corrosion, owing to the alkaline nature of the lime binder in the hempcrete. Ideally, the two materials should be kept separate. Where the frame does need to interact with the hempcrete, it should be protected with a painted coating; the same applies to any fixings, which could alternatively be made from an alkali-resistant metal such as stainless steel.
- **Glulam.** If the glulam is to be encased or partly encased in the hempcrete, check with the manufacturer that the wet environment of the freshly placed hempcrete will not have an adverse effect on the glues in the beams. If there is a risk that it will, a protective coating may need to be applied.
- **Oak, and other hardwoods.** A traditional pegged hardwood frame does not have any of the issues of the above materials in its interaction with hempcrete, but there are limits to the size of frame you can create in this way.

Steel and glulam interact in the frame of a large superstore built using Lime Technology's Hembuild® panels.

An oak frame, with hempcrete walls under construction.

A steel supporting frame in a hempcrete community centre.

Glulam provides a more sustainable alternative to massive hardwood beams or steel joists.

Smaller buildings

Buildings of up to three storeys can be constructed using a simple studwork frame, and this is the usual approach taken for any hempcrete building up to the size of a large house. The discussion in the rest of this chapter relates to the timber frame for a hempcrete building on this scale. The frame can usually be constructed from untreated European softwood, which is a natural and sustainable resource.

For very simple builds, such as small extensions or simple one- or two-storey buildings, those who have the necessary skills may wish to design their own frame. The stud frame required for these types of building is simple enough that building control inspectors will be familiar with its structural limits. Whether you are undertaking the design of the structural frame yourself or whether it is being done by an experienced engineer, we advise that the standards of construction set out by the Timber Research and

Double frame

In situations where external cladding *and* internal permanent shuttering board is desired, a double-frame design can be used. This comprises an exposed frame flush with the internal face, which acts as the structural frame, into which the permanent shuttering is fixed, and a second, non-load-bearing frame flush with the external face to provide a fix for the cladding. The non-load-bearing cladding frame is usually of smaller-section timber.

The two frames are connected using rectangles of OSB or plywood screwed into the side face of each stud so that the edge sits flush with the external face of the timbers. The dimensions of the rectangles of board varies according to the design and width of the wall, and they are fixed at regular centres going up the studs, as specified in the frame design.

The double frame is quicker and easier to shutter than a single frame, but takes longer to build and is not the ideal frame design, because of the potential for thermal bridging and air leakage caused by the connecting squares of OSB.

Elements of the frame

The following discussion of individual elements of the frame incorporates the basic elements, as described previously, together with some other elements that are not necessarily present in all frames.

Sole plate

The sole plate is positioned on the plinth above the DPC, or, in some circumstances (if the plinth is narrower than the foundation), sits inside the plinth and bears directly on the foundation blocks. It is either fixed directly to the surface it bears on

to, or sits on the plinth and is tied back to the foundation or floor slab using long metal straps (bent straps). These straps are usually available only in galvanized rather than stainless steel, and so need painting after fixing to protect them from corrosion.

The function of the sole plate is to provide a solid fix for the studs, and also to stop them moving independently of one another. The sole plate should be joined continuously along its whole length, which sometimes means doubling up timbers and overlapping to ensure a strong join between two lengths of timber.

The section size of the sole plate timber is dependent on the section size of the studs that are fixed to it, with the same-section timber being used for both.

Studs

The function of the studs is to take the load of the roof and upper floors down to the foundations, and to provide support for the hempcrete. Studs

Metal straps are used to secure the sole plate.

are fixed to the sole plate with skewed stainless-steel nails or screws (or hot-dip galvanized nails), or stainless-steel angle brackets, and extend to the wall plate at the floor or roof above (see Figure 5). They are spaced at regular centres, usually 400mm or 600mm depending on the specifications of the frame.

The section size of the studs depends on the load they are taking, the height they extend to and other aspects of the frame design, such as whether a particular spacing is desired, for example to tie in with floor joist spacing. A 100mm x 50mm stud is usual for single-storey buildings, while a two-storey building normally requires 150mm x 50mm studs.

Multiple or reinforced studs

These are studs fixed together to make larger-section timbers, which are positioned under areas of greater loading where additional strength is required. For example, they may be used to support a large beam or a lintel (see Figure 6 overleaf), or at the corner of a building.

They are fixed together using nails, screws or even coach bolts if they are very large, and may require additional fixings to the sole or wall plates via stainless-steel angle brackets.

Wall plate

The wall plate runs along the top of the studs and is fixed on to them from above, using two stainless-steel nails or screws per stud. The function of the wall plate is to provide a fix for the top of the studs, stopping them moving independently from each other, and also provides a solid surface for the roof timbers and/or upper floor joists to bear on. The wall plate is usually identical in construction to the sole plate and, again, should be continuously joined along its whole length.

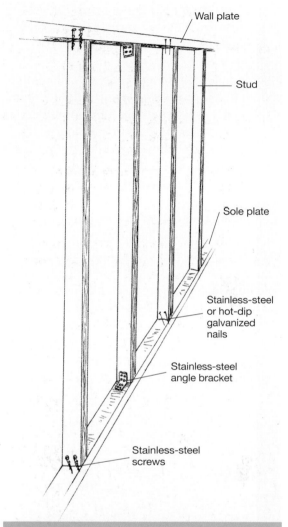

Figure 5. The basic frame, showing alternative methods of fixing the studs to the sole and wall plates.

Floor and ceiling joists

Floor and ceiling joists obviously create the upper floors, but within the frame they also serve to tie opposite walls to one another, creating a stable box section, which prevents walls from leaning outwards. These joists can run in only one direction and therefore don't connect the walls that run parallel to them. Those walls are tied in to the floor or ceiling construction using bent straps (see Figure 7 opposite), which extend from the wall across two or three joists at regular centres, usually every other stud.

Floor joist section size depends on the loading and the distance they must span. For longer spans, an engineered joist (see page 158) is normally used. Joists can sit on the wall plate and be fixed down into it from above, or be hung from the wall plate on metal joist hangers (see Figure 8 opposite). The first solution is generally preferred, since it leaves less metal exposed to the hempcrete, and in any case stainless-steel joist hangers are costly.

Framework for openings

Openings in the wall are created by horizontal and vertical timbers within the framework to define the opening and provide a fix for door and window frames. These timbers are also used at the shuttering stage, as the shuttering to form the opening is fixed to them.

Framework for openings in the wall may sit in the same plane as the structural frame or be offset from it, closer to either the internal or external surface, as required.

Sections of the timber framework around openings are shown in Figures 36 and 37 (pages 337 and 339).

Lintels

Lintels are the timbers that span the top of windows, doors and other openings in the wall. Their specification depends on the size of span across the opening and whether they are also supporting a floor joist, and thus how much load they are taking.

In its most simple form – over doors or windows that are not supporting upper floor loads – the lintel may simply be part of the framework for the door or window, and consist only of a

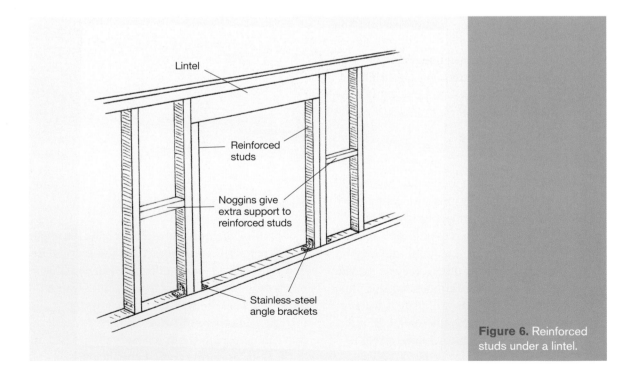

Lintel

Reinforced studs

Noggins give extra support to reinforced studs

Stainless-steel angle brackets

Figure 6. Reinforced studs under a lintel.

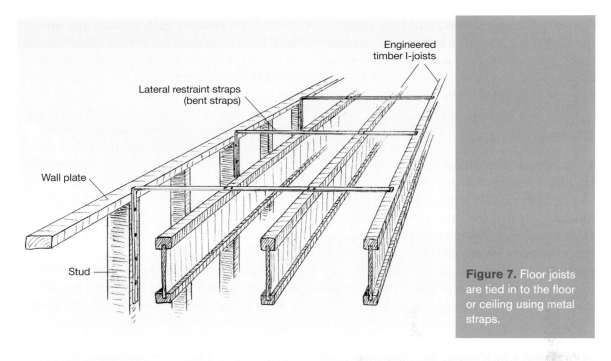

Engineered
timber I-joists

Lateral restraint straps
(bent straps)

Wall plate

Stud

Figure 7. Floor joists
are tied in to the floor
or ceiling using metal
straps.

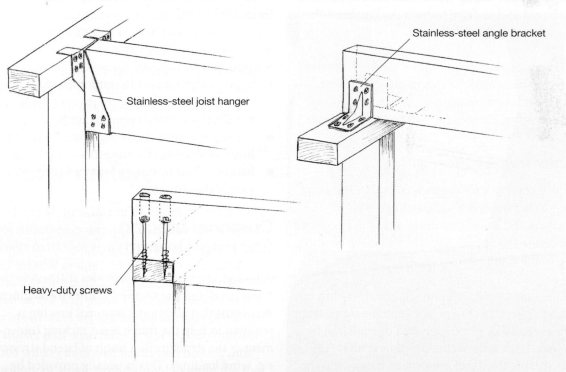

Stainless-steel angle bracket

Stainless-steel joist hanger

Heavy-duty screws

Figure 8. Floor joists may either bear on the wall plate or be hung from it using joist hangers. If bearing on the wall plate, fixings may be angle brackets or screws.

For the most part, roof design options for a hempcrete building are no different from those for conventional construction. The only difference in roof structure that often applies to hempcrete buildings is a long roof overhang at the eaves for additional rain protection, which is achieved using longer rafters or additional bolt-on rafter ends, as described in Chapter 22.

Practical considerations relating to the frame

At the frame design stage, it helps to think about the later stages of the build and to give some thought to practical issues that are likely to come up in the construction process. A few examples are outlined below.

When constructing a frame, consider the easiest way to go about it. Decide in advance how each section of the frame will be constructed and erected on-site, and whether temporary supports are needed for one section before the next can be put in place. Temporary braces can easily be screwed into the frame and removed again with no concern about marking the frame timbers, as these will be buried in the wall or covered with plaster, render or cladding.

It is possible to prefabricate large sections of the frame on the floor and lift them into place to be fixed in one go rather than lifting each individual timber into place, thus eliminating the need to both support and fix each timber at the same time.

Another approach that can make framing faster, if your design allows it, is to prefabricate the frame in sections in the workshop and bring it to site to be assembled. The main advantage of this is that the frame construction can be done more efficiently and quickly in a dedicated workshop, with machine tools, than among the relative chaos of a busy building site. It also means that

the framers spend less time on-site, and it allows the frame to be quickly assembled as soon as the plinth is ready.

As noted earlier, it is important when designing the frame to consider the process of placing hempcrete evenly around the timbers, and how this will be affected by the frame design. As far as possible, the design should avoid creating areas that are hard to reach. This is why it is useful to minimize the use of horizontal timbers.

Think about other ways in which the frame design can be used to reduce work later in the build. For example, consider how the frame interacts with the erection of temporary shuttering around it. Shuttering boards are held off the frame at the correct distance by long screws and specially cut spacers made from plastic tube. So

Design the frame so that shuttering spacers are the same length on both sides.

you can save work by having your frame positioned *exactly* centrally within the wall – meaning that the spacers needed for the external and the internal shuttering will be exactly the same length, which greatly simplifies the job. Moreover, if the frame is offset from the centre only slightly, the different-sized spacers needed to set the shuttering on each side of the frame will differ only slightly in length. The problem with this is that two very similar-sized lengths of plastic tube can look almost identical when they are chucked in together in a bucket covered in dried lime. Given that on a large building you could easily need upwards of 50 or 60 spacers for each side of the frame, you can imagine the amount of time you can waste sorting out which ones you need – as well as undoing shuttering that is not held firmly, or is out of true, because the wrong spacer has been used.

Remember that even though the frame is not on display, it is important that it is constructed with the utmost accuracy (see Chapter 8). The shuttering that will form the faces of the final hempcrete walls is constructed off this frame, so timbers must be placed accurately within the frame so that it's easy to fix the shuttering in the right place, first time, without too much fiddling about adjusting spacers and fixing then re-fixing the boards.

The exact placement of the frame leads to the correct placement of the shuttering, which leads to flat, straight and plumb walls, which makes plastering easier. The plastered wall is the only thing you will eventually see, but in a hempcrete building the quality of the plaster finish starts with the frame, so take time from the very start to make sure everything is as good as it can be!

Accuracy at the framing stage makes applying render and plaster finishes easier, and allows the formation of very straight walls, as demonstrated in the clean lines of this garden office building.

CHAPTER FOURTEEN

Shuttering

In the cast-in-situ method, hempcrete is cast around a frame on-site, with the aid of shuttering boards, which keep the freshly mixed hempcrete in place while it sets. Shuttering may be either temporary or permanent.

Temporary shuttering boards

Temporary shuttering is erected to create the mould for the internal and external faces of a hempcrete wall. Once the hempcrete has been cast and has taken its initial set, the shuttering is removed and moved on around the frame so the next section can be cast. Temporary shuttering is usually either attached to the frame or built out from it using long screws and spacers.

Shuttering boards need to be stiff enough to hold their shape without lots of extra reinforcement, yet light enough to handle and carry easily in large sheets, and hard-wearing enough to be used again and again. There is no definitive rule about what type of board should be used, but the following criteria should be considered. No one board will meet all the criteria, but with a bit of

thought a good compromise can be achieved. The board should be:

- hard-wearing
- cheap
- lightweight
- stiff/flexible as required (e.g. you may want to cast a curved section of wall or recess)
- easy to fix (to the frame and to neighbouring boards)
- easy to mark on and cut
- able to be recycled elsewhere in the building project once the hempcrete is cast (e.g. incorporated into the building as floor or roof boarding, or used to board out a shed or workshop).

You will probably also want to consider whether the board is:

- low in embodied energy
- made from natural materials or synthetic materials
- made from materials from a sustainable source.

Early on at Hemp-LimeConstruct we experimented with using chipboard flooring, which easily fits together with the next board as it comes with tongue-and-grooved edges, and this reduces the amount of fixings required and

Left: For this building the whole of the outside was shuttered in advance, then the hempcrete placed from inside.

- Fix the whole run around the building, with corners shuttered as described opposite. As each board is put up, join it to the previous one with wooden straps or ties, as described on the previous page.
- Once a few consecutive boards are up, check them for the line vertically and horizontally (with a long spirit level) and adjust the screws as necessary.
- Once the void is filled with hempcrete (see Chapters 15 and 16), remove each top screw in turn, taking out the spacer tube and replacing the screw. The face of the freshly cast hempcrete now becomes the 'spacer', forming a guide to how far the board needs to be pulled back in. Be careful not to over-tighten the screw, as the hempcrete is still wet and easily pushed out of shape.
- Fix three or four straps to the top of each board, projecting upwards, and place the next lift of shuttering above the first one, fixing each board with the straps at the bottom and a single spacer and screw on each stud at the top edge of the board.
- Continue the run, fixing and checking the line as described above.
- Aim to take the shuttering boards straight across any small openings in the wall (e.g. windows) that aren't needed for access to the building during the hempcrete-placing stage. This makes the process of keeping the wall straight much easier (see page 185).

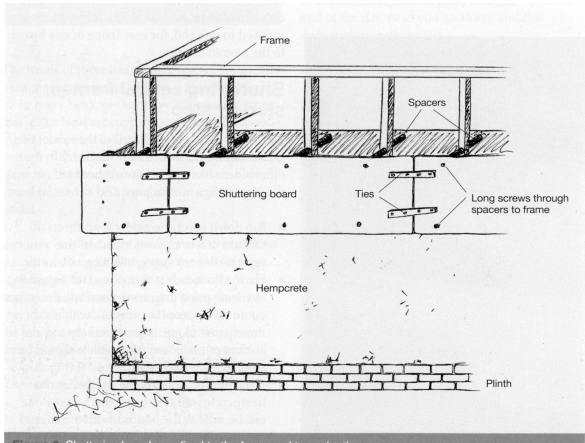

Figure 9. Shuttering boards are fixed to the frame and to each other.

As described earlier, the shuttering goes up in 600mm lifts on the side where people are working to fill it, but on the other side it can be created from full boards. In the case of a central frame, however, full boards are much harder to put up, since they are heavier and therefore more difficult to manhandle and to hold in place while the fixings are going in.

They have the added disadvantage that they require three rows of fixings, and – while this means fewer fixings overall, compared to two half-sized boards with two rows of fixings each – the bottom two rows of fixings are hard to fit, because it's impossible to reach around a full board to hold the spacers in place. Therefore one person needs to be positioned on the other side of the wall with the spacers, so that when the screw is started through the board, they can hold the spacer in place and guide the screw to the correct place on the frame as it is driven in.

Putting up shuttering on a central frame is always a two-person job, with one person to support the board in place (usually with the board resting on the section of shuttering below it, or on the ground) while the first two or three fixings are put in. Once there are enough fixings in to hold the board loosely in place, one person can put the others in, while the other person attaches two or three straps to tie the board to the end of the previous board in the line, and three or four to tie it to the board below, if these were not already added to the top of that board.

When a complete run of shuttering is fixed in place, take a long straight edge, and the longest level possible in the space, and check that the new shuttering is plumb vertically, and forming a straight continuous surface horizontally with the last board that you know to be true. If any adjustments need to be made this can usually just be done by loosening off the screws which go through the spacers or pulling them in slightly.

Shuttering at corners

The erection of shuttering at corners in a way that will create a straight, square, plumb corner is of course inherently more difficult than erecting straight runs of shuttering. It is also an area with very little tolerance for error, as any small mistake will result in a highly visible disturbance to the straight lines of the building. That said, straight lines in a building are not necessarily what everyone is aiming for – but if they *are* important to you, the following methods should help you to achieve them.

Corners externally:
- Cut the last board of the run so it over-sails the corner, fix it, check it for plumb and for straightness against neighbouring boards, and adjust as necessary.
- Start the next run of shuttering by placing the next board at right angles, butting up against the one that over-sails.
- With this board fixed loosely in place (with screws and spacers), use an appropriately sized square to check that the second board is at 90 degrees to the first, paying attention to top and bottom of the board (the middle should follow, in theory!).
- When you are happy with the position of both, the boards are held in place by one person while the other screws through the face of the over-sailing board into the end of the second board using 50mm screws. Screwing into the edge of an 11mm OSB sheet does not make for a very strong fixing, so use as many screws as are needed to keep it firmly in place.
- Tighten the fixings on the first board in the new run, and check it again for plumb and square at the corner (using a level on both faces, and a square outside and inside the corner formed by the boards), adjusting as necessary.
- Continue with the new run, working off this first board.

Shuttering for openings in the wall.

adapted to create openings that go part-way through the wall. These might include, for example, alcoves, shelves or cupboards that are inset into the finished wall and are partly or completely formed by the casting of the hempcrete itself, in whatever form pleases you. Obviously there is a limit to the depth to which the alcove can extend without significantly reducing the insulation provided by, and the strength of, the hempcrete that remains.

A small-section frame detail around the alcove opening can be useful in this situation, not only for fixing the temporary shuttering but also for ease of fixing the finished lining, for example wooden shelves, a lining around the recess or a cupboard door, into the opening.

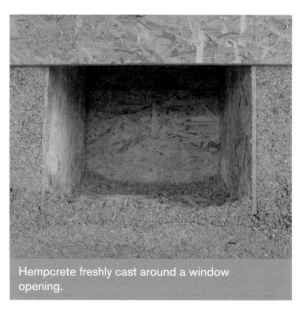

Hempcrete freshly cast around a window opening.

Window openings with shuttering partially removed.

Curved walls

It is possible to cast curved hempcrete walls, by constructing a curved frame and holding shuttering boards off the frame with spacers in the usual way. In this situation, experimentation is needed to find boards that bend enough to create the desired curve without breaking, while remaining stiff enough to hold their shape in a smooth, continuous curve. The obvious choice for this is a thin ply, but a 15mm wood wool board curves pretty well too.

A note on shuttering historic frames

Shuttering for casting hempcrete in the context of historic restoration work is potentially more complex than on a new build, and has to be designed on a job-by-job basis according to the particular needs of the application. The shuttering design itself is not necessarily difficult, but

A curved hempcrete wall enclosing a shower. (The inside face is finished with tadelakt – lime-based and waterproof – see page 249). *Image: Margot Voase*

Hempcrete can be shaped with a nail or grid float after casting.

the materials used for straight walls in new-build applications may not be suitable.

As anyone who has lived in a house that was built before the late nineteenth century knows, walls in old buildings are never straight. For this reason, when setting up shuttering, especially on ancient timber frames, a more pliable bendy material is often required, so that the shuttering can easily follow the line of the wall. However, bendy shuttering is only so bendy; where necessary, a 'best fit' shuttering solution is constructed and the rough or unwanted edges of the cast material can be scratched back and shaped with a nail or grid float after the shuttering is struck.

Remember that historic timber frames are often left visible in the face of the hempcrete wall, and therefore it might not be appropriate to screw shuttering boards directly into the face of the timbers. This problem can be overcome by screwing battens centrally to the inside faces of

Shuttering and casting around ancient timber frames, such as in this Tudor barn, can be a slow business.

the timbers (i.e. the side of each timber that faces in towards the hempcrete panel) and screwing through the shuttering boards into the batten (without spacers so the shuttering is pulled in tight to the shape of the frame). The battens are buried below the surface of the cast hempcrete, and the screws are easily removed when striking the shuttering.

Bear in mind that in historic work the casting of hempcrete is likely to be a slower and more intricate process, with lots of existing wall structures that get in the way of casting and very much need to be 'worked around' rather than removed. The final wall thickness, if determined by the historic frame, is often a lot thinner than in a new build, which makes it a more fiddly business to spread hempcrete evenly in the void. For this reason the wall may be cast in smaller lifts, and the height of each shuttering board needs to be adapted accordingly.

For more information on the use of hempcrete in historic buildings, see Chapter 20.

Plastic shuttering

Contractors may be tempted to invest in one of the modular reusable formwork systems that are used for casting concrete. These are often made from recycled plastic, and are available in a range of designs, including rounded pieces for casting columns.

This option might at first seem a sensible route to go down, as the modules are very easy and quick to assemble, keep the wall very straight and reduce wastage (if you are throwing away your OSB when it gets too tatty, or has been cut up too small).

On the other hand, they are expensive to buy – so much so that you would only begin to see a

saving if you were doing a lot of large builds. They also need cleaning well after each build, so that any bits of dried hempcrete on their surface will not stop them producing a clean, straight wall on the next build, which is an extra job. Lastly, they are much bulkier than thin wooden boards, and you would need to have lots of them, and therefore would have to think about where to store them and how to transport them.

As described earlier in this chapter, at Hemp-LimeConstruct we have concluded that the most practical and economical solution is to use OSB, since the finished wall it makes is no worse or better than one formed with plastic shuttering. OSB is cheap, readily available and easy to store, it's made (mostly) from natural materials, and you can reuse it many times over as shuttering, and then use it in the building or build a shed with it afterwards.

Plastic reusable shuttering. *Image: The Limecrete Company*

CHAPTER FIFTEEN

Mixing hempcrete

This and the next chapter are arguably the most important of the whole book. We would very much like you to read them more than once – indeed, as many times as necessary in order to absorb all the information well. For a 'belt and braces' approach you could write down the key steps, and have them handy the first few times you make a hemp–lime mix.

Mixing hempcrete is easy to get wrong when you are not practised at it, and there is not much margin for error. Getting the mix even slightly wrong can have serious negative effects on the finished wall. For this reason we recommend that you not only follow the binder product manufacturer's guidelines to the letter, but also pay close attention to the principles and method described here. It goes without saying that if you have not done it before, you need to make up and cast some test mixes of hempcrete before starting on a real project. The construction of a section of test wall enables you check and adjust both your mixing and placing techniques.

Basic principles

The first thing to say is that personal protective equipment (PPE) is extremely important when mixing hempcrete. If you attempt to do it without the correct equipment you *will* get chemical burns from the lime in the binder. Lime burns are painful, and can cause serious health problems in the case of getting it in your eyes (permanent loss of vision), or airway (acute irritation and/or chronic respiratory problems).

Read Chapter 9 (Health and safety) very carefully, and make sure you put it into practice. You are *not* the exception to the rule. Even if you are a bloke. What's more, if you are employing people to do this work, you (and also they) have a duty to ensure that safe working practices are followed, that the correct PPE is worn or used, and that proper training and instruction is given to anyone carrying out this work. You also need to keep a first aid kit handy with the correct items in it, the most crucial of which is eye wash.

Left: Hemp-LimeConstruct's pan mixer has been specially designed for mixing hempcrete.

The key principle to remember when mixing hempcrete is the importance of the mix including all the constituent materials – hemp shiv, binder and water – in the right proportions. For those used to working with cement mortars and concrete, this may seem strange: with these, there is much more leeway for the mix having slightly too much, or too little, cement in it. The gauging of the cement binder to the required sand and/or gravel is usually done quite roughly with a spade or shovel. And the worst consequence of a cement mix that is too wet or too dry is likely to be your boss grumbling about its workability, rather than it having a serious effect on the building's structural integrity. Not so with hempcrete. Because the aggregate in hempcrete is a plant material, and because the desired consistency is a light and fluffy (low-density) mix, with every piece of hemp shiv coated in binder, exactly the correct proportions must be added.

Just the right amount of water

Water can be measured into the mix in different ways, and in some cases small adjustments may

The right consistency is not too wet and not too dry.

have to be made, usually by eye, to account for variations in the weather conditions or the hemp shiv being wet. However, as a general principle, precision about the proportions of ingredients is especially important in terms of the amount of water added. The consequences of either too much or too little water can be serious.

Not enough water

The hemp shiv arrives at the site in a dry state, and is carefully stored on-site to ensure that it does not get wet. Failure to do this can result in it getting damp and mouldy and beginning to rot before it can be protected by being mixed with the binder. The lime-based binder is hydraulic, meaning that it requires water for the chemical reaction by which it sets (see Chapter 3). A certain amount of binder requires a certain amount of water to allow it to set fully. However, because the hemp is dry, it naturally absorbs water when it comes into contact with it, and so when the hemp and the binder are combined with water, the two ingredients are actually in competition for the water in the mix.

The effect of not enough water in the mix is that the water-hungry dry hemp soaks up water needed by the binder, leaving some of the binder in a powdery or semi-powdery state. Result: your wall fails to achieve full strength and is at risk of collapse.

Too much water

Well that's easy then, you might say – simply over-compensate and make sure there's plenty of water in the mix: a little bit over won't hurt – better safe than sorry? Well . . . no. Unfortunately, because we are using a natural plant aggregate, one of the risks is that excessive or extended exposure to water can cause the hemp to rot inside the wall. This is in fact an unlikely result, because the hemp's natural properties allow it to absorb water and release it again without any damage – that's why it is so suitable for a breathable wall system. However, any prolonged period of exposure to excess water is to be avoided.

In addition, we need to be sure that the water we introduce at the mixing stage is able to dry out of the wall within a reasonable time period, so that finishes can be applied. Although the wet finishes used will be vapour permeable, they restrict the speed at which moisture leaves the wall, because compared with the bare hempcrete they have a less open structure and therefore a smaller surface area from which water can evaporate.

This second effect of having too much water in the mix is probably the more serious. In our experience this is often a major bone of contention on building sites, as any perceived delay in applying finishes is assumed to be costing time and money. In practice, this is not always true, and it would make sense in most cases to leave at least one side of the hempcrete walls without finishes for the longest possible period, to be certain that they have dried out to their optimum 'resting moisture content' as quickly as possible. (see Chapter 16, page 215). Frequently, however, there is pressure to complete the wall as quickly as possible, and assuming this to be the case, you really don't want to be introducing any more water into the mix than is absolutely necessary.

Another effect of too much water is that the weight of the mix increases, causing the placed hempcrete to compact more closely, resulting in a denser material and a reduced insulation value. When excess water in the mix reaches a critical level, the wall could collapse when the shuttering is removed, because the weight of the wall is too great for the strength of the initial set of the binder to hold. For this reason, as well as those already described, we really cannot overemphasize the importance of using the correct amount of water.

Pan mixer opened to show paddles and hempcrete mix. (Note: *never* use a pan mixer with the lid open!)

Who is doing your mixing?

It goes without saying that the best way of ensuring correct proportions in the mix is by accurate measurement every time. This is partly addressed in the mixing method (described on the following pages), but you can't avoid the fact that some people in your team are going to be more suited to this job than others, and getting the right person (or people) on the mixer will be

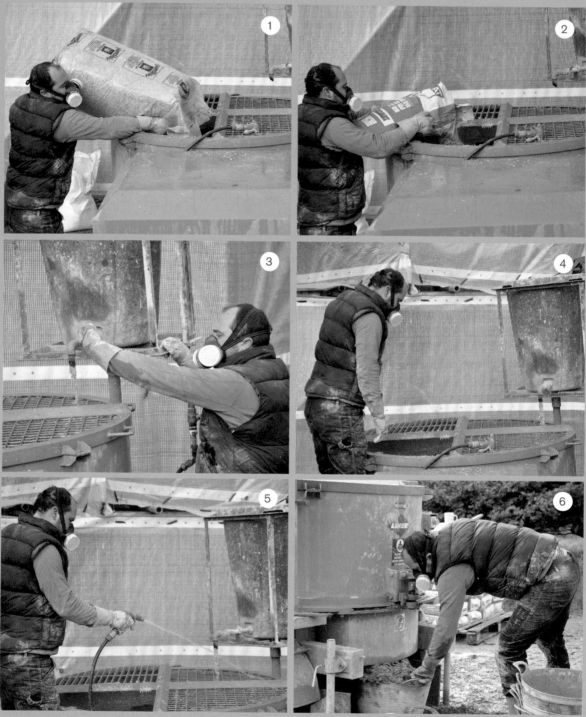

Mixing in a pan mixer: step by step

1. Add the hemp shiv.
2. Add the binder and start the mixer.
3. Add water.
4. Mix until well combined.
5. Adjust the amount of water by spraying down.
6. Empty the hempcrete into tubs.

The mixing process is described in more detail below right, but these are the basic principles:

- The binder should be well mixed with water first (to ensure sufficient water is taken up by the binder).
- The hemp shiv is then added, then, after more mixing, the rest of the water, and all materials are evenly combined to achieve a consistent finished mix. However, depending on the individual mixer you are using, it may not be possible to add the materials in this order. Sometimes the pan is not completely water-tight (though it should be), so, if combining water and binder first, some may escape. Also, when the hemp is added last it sometimes clogs together as it goes in and doesn't mix easily. In either of these situations, it is acceptable to mix the hemp and dry binder thoroughly first and then add the water (as shown in the photos opposite), although care should be taken to distribute the water evenly through the mix.
- The mixing should be long enough to achieve an evenly blended material and no longer. This time can vary with the materials, but is easily judged by eye, since it is quite easy to see into the mixer.
- Once mixed, the wet hempcrete should be placed immediately, or within the time recommended by the manufacturer (depending on which binder you are using), and this time limit should be strictly observed.
- If you are not able to use the mix straight away, the hempcrete should be stored in a tub in *clean* and *dry* conditions so that no contaminants and, most importantly, *no extra water* gets introduced to the mix.

For our Hemp-LimeConstruct Example Mix, with our imaginary HLC Binder (see page 195 – do *not* use these proportions in your real building), we need to make a mix of the proportions 4:1:1 (hemp:binder:water). In terms of what will easily fit in our large pan mixer, this equates to (per mix):

360 litres of hemp (2 x 180-litre bales) + 90kg of 'HLC Binder' (3 x 30kg bags) + 90 litres of water

Mixing process

The method will be prescribed by the binder manufacturer and/or supplier, but in general the process of mixing is as follows:

- Put on goggles and mask (or full-face mask), waterproof clothing, gloves and waterproof boots.
- Pre-fill either marked tubs or buckets, or a gravity-release tank with the required amount of water, so that the precise amount is already measured out before mixing begins.
- Add approximately half of the water to the mixer and let its paddle move the water around to collect any leftover residue from the last mix.
- Add the binder, opening the bags and emptying them through the centre of the mixer grille in such a way as to minimize the dust released into the air above the mixer. Ensure that *all* the binder has been removed from each bag.
- Allow the water and binder to mix to a smooth slurry, and meanwhile use a broom to knock in any binder that got caught on the grille of the mixer. Dispose of binder bags safely in a way that limits the amount of binder dust spreading around the working area.
- Open the bales and dump the hemp (a bale at a time) in the centre of the grille. Use a broom to work the hemp down through the grille into the mixer, being careful not to brush it off the side of the mixer. Again – although it is less of an irritant than the binder dust – take care to ensure that minimal hemp dust is released into the air.
- Allow the hemp to mix into the slurry, until a dry but fairly consistent blend is achieved. At this point, or if no further mixing can occur due to a lack of water, add the remaining

water. As far as possible, try to add the water in such a way that it is evenly spread around the mixer rather than being concentrated in one part of the mix.

- Note: if using the alternative method described in the 'basic principles' on the previous page, in which the hemp and binder are mixed dry before water is added, follow these steps: With the mixer off, put the hemp in, then add the binder, distributing it as evenly as possible. Start the mixer and run it until the hemp and binder have integrated to become a consistent blend. With the mixer still turning, add the water, ensuring it is distributed as evenly as possible throughout the mix as it goes in.

- Allow the mixer to do its job, until the mix seems to have the right consistency and looks completely even. You are trying to achieve a light, fluffy mix and to avoid it 'balling up' (small balled lumps in the mix, which mostly consist of binder).

- If you see balled lumps appearing, stop mixing immediately. There is not much you can do to get rid of them again, but make the placing team aware of it and ask them to discard any balls they see in the tub as they are placing the hempcrete.

- To test the consistency, drop a little of the mix out into a tub, pull out a handful, and squeeze it into a ball with both gloved hands. No liquid should drip out. Hold the ball in the palm of one hand and push a finger in. If the ball crumbles, it is too dry; if your finger just 'squidges' into the ball, it is too wet. If it splits in two, it is just right (see photos below).

- If it is not right, throw the pile in the tub back into the mixer and adjust the consistency by adding a little more water or hemp as required, and adjust the amount of water going into the next mix if necessary.

- When the proper consistency is achieved, leave the mixer turning, place a Gorilla Tub under the chute and fill it by opening and closing the chute door with the lever. Repeat until the mixer is completely emptied. It's easiest to have one person mixing and one working the chute (see opposite).

- Try to avoid any delay in emptying the mixer. If delay is unavoidable, stop it turning, as over-mixing will cause the mix to start balling up.

Testing the mix
1. Squeeze some hempcrete into a ball.
2. Push a gloved finger into it.
3. If it breaks cleanly, then it is ready.

■ When the mixer is empty, ensure that the chute is properly closed (watch out for pieces of hemp getting stuck in the chute door), and use some of the next mix's allowance of water to wash any bits from the sides. Then you are ready to either start a new mix or turn off the machine if another mix is not needed immediately.

■ Any spilt bits of hempcrete by the mixer, or bits dropped outside the shuttering, can be swept up – as long as they are uncontaminated by other materials – and added to the next mix to replace fresh hemp shiv. However, this must not make up more than 10 per cent of a whole batch. You are more likely to be able to reuse dropped hempcrete mix if you deploy clean OSB sheets or similar at the bottom of the shuttering where people are placing, and underneath the mixer chute.

With a pan mixer, the mixing team typically includes two people: one organizing the materials and feeding the mixer, and another below operating the chute, i.e. filling tubs and handing them to a team of people ferrying them to the 'tampers', who are placing the mix into the shuttering. Once each tub has been emptied into the shuttering, the ferrying team returns the empty tubs to the mixer to be refilled and brought back, or to wait for the next mix. However, it may be possible for one experienced person to operate the mixer alone, and this is certainly possible if it has a built-in water tank, automating the process of measuring and adding water.

The quantity produced with an 800-litre pan mixer will be (depending on the exact quantities used) around 0.4m^3 per mix. This of course takes a while to produce, given the mixing time, the physical work involved in feeding the mixer, and the time taken to empty it into 30-litre or 45-litre Gorilla Tubs (the 30-litre type is more useful, as everyone can carry them) before a new batch can be started. It is important at first to

give your mixing team time to get used to the process, so start slowly: with hempcrete, getting it right is always more important than getting it done quickly.

Once you are used to the process, you will work out a general rule of thumb for the quantity of mix you can knock out in a normal working day, but remember that the mixing speed depends on a lot of other factors as well, such as other people needing to use the telehandler, and the distance that the ferrying team have to walk to empty the tubs. Ultimately, of course, the mixing speed has to match the speed that the placing team can work at, which can vary around different parts of the frame (placing slows down around corners and openings, and also relies on the shuttering being moved on ahead), as you don't want tubs of mix sitting around drying out or getting rained on. In any case, you need the tubs back to keep emptying the mixer, unless you want to invest in more than 25 tubs.

We return to this subject in Chapter 19, but for now, the point to remember is that finding the balance between the speed of mixing, ferrying and placing hempcrete is important. The balance needs to be constantly monitored and readjusted as necessary, in order that the operation is run efficiently and maximum productivity is achieved.

Mixing in a bell mixer

For applications where placement of the hempcrete will be slower (e.g. fiddly historic building work, or work with complicated or narrow shuttering), or where the total amount placed is smaller (e.g. a small building or extension), it is possible to mix hempcrete in a standard bell (or 'drum') concrete mixer. This is not recommended for large new-build applications, as you will not be able to produce enough to keep up with the speed of placement work.

Mixing in a bell mixer: step by step

1. Measure out the ingredients.
2. Add the binder.
3. Add the hemp shiv.

4. Mix until well combined, but no longer.
5. Turn half-turns and clean the sides.
6. When the mix is ready, empty it into tubs.

- At this point, watch the mix as it turns, because it now becomes important not to leave the hempcrete mixing any longer than is necessary, otherwise it starts to form into balls, preventing an even mixing of the hemp and binder (see below).

- When the third quarter of hemp is evenly distributed, stop the mixer and reach in with a gloved hand to pull off any mix that is stuck to the back of the drum. Then start-stop it once to turn the bell through 180 degrees, and repeat the process.

- Start the mixer and add the last bucket of hemp. Watch it closely and don't let the mixer turn any more than it needs for all the hemp to blend in and become a consistent colour (approximately 1 minute).

- Stop the mixer and pull any mix off the back as before, then let it turn for a further 10 seconds.

- Stop the mixer. Your mix should now be ready.

- To test whether it is the correct consistency, use the ball-and-finger test described on page 200. You are trying to achieve a light fluffy mix, whilst avoiding the mix balling up (see also page 200). Balling up is best avoided by keeping the mixing time to the absolute minimum necessary. In a bell mixer a perfectly good mix will turn to balls if you leave it going too long. Keeping the mixer in the most horizontal position possible after you add the second half of the hemp also helps to prevent balling.

- If the consistency is not right, adjust it by adding a tiny bit more water or hemp *and* binder accordingly, turning the mixer only the absolute minimum for all the hemp to mix in consistently, as described above. If necessary, adjust the amount of water you will use for subsequent mixes (never adjust the ratios of binder to hemp).

- You may find that on hot dry days the loose hemp shiv is exceptionally dry and a little bit more water is required, but always stay conscious of the fact that you want to add only the very minimum of extra water to the mix, as every unnecessary litre you put in will increase the drying time of the finished wall.

- Sometimes, for no apparent reason, you get a mix that isn't combining as well as it should. If this happens, stop the mixer and thrash the hempcrete around with your hand a bit. Turn the mixer 180 degrees and do the same again. Let it turn a bit more, and repeat this step until the mix looks correct.

- When you have tipped your mix into Gorilla Tubs, throw some water in and clean the mixer out, so that you don't get bits drying on the side of the bell and affecting the success of the next batch.

- Any spilt bits of mix that are still 'clean' (uncontaminated) can be swept up and put back into the next mix, as long as the guidelines described on page 199 for using a pan mixer are followed.

The bell mixer method may sound a very long-winded and time-consuming way of mixing, but once you get the hang of it it's possible for one person to keep two mixers constantly running – as long as the materials are close to hand and you have a good pressure on your water supply. You can get into a rhythm so that while the mixer is turning you're filling buckets ready for the next batch. One mix takes about 5-10 minutes, and it is possible to mix around 2-2.5m³ of hempcrete in a day.

The point noted on page 201 about finding a balance between the speed of mixing, ferrying and placing hempcrete, applies here. But suffice to say that if you have chosen to use a bell mixer, you have already decided that you are doing a job where large quantities of mix are not required quickly.

CHAPTER SIXTEEN

Placing hempcrete

The process of placing hempcrete is a deceptively simple one: while it is true that the basic method is straightforward, it does require some skill and an understanding of the important qualities of the finished cast hempcrete material. Moreover, it demands knowledge of the effects that differences in placing technique can have on these key qualities.

Before we look at the method itself, it is worth briefly discussing the principles involved, as a good understanding of these underpins successful technique.

Basic principles

When placing hempcrete, the aim should be:
- to fill the whole of each shuttered void evenly and consistently with hempcrete, with no gaps, and achieve a good proximity of the placed hempcrete to the frame timbers
- to achieve a consistent, low-density cast

material: strong enough to hold its own shape, but trapping the maximum amount of air inside it to provide as much insulation as possible.

In order to achieve a consistent, low-density material *it is important to avoid over-compacting*. At Hemp-LimeConstruct we very rarely use a 'tamping stick' (or any other tool) to compact the hempcrete in the void, as we have found that this encourages people to over-compact it. This is because if your hand is not in contact with the hempcrete itself, you cannot feel how much it has been compacted.

There are two exceptions to this rule. A stick can be used if needed to extend your reach into an area of the void that is difficult or too far away to reach by hand. In this situation the stick should be used, as far as is possible, in a way that replicates the motion of your gloved hand. A stick is also used when tamping *is* required – which is usually in areas of thin coverage over frame timbers, where more compaction is needed to

Left: Placing freshly mixed hempcrete into shuttering.

The (in)famous tamping stick: never used unless *absolutely* necessary.

give increased structural integrity, at the cost of a little loss of insulation. A short length of roofing batten is generally used for this purpose, but occasionally a plate is screwed on to the bottom to create a tamping stick with a slightly wider area.

The density of the cast hempcrete is the key quality of the finished material, and can be thought of as a continuum, as illustrated in Figure 11 below.

The finished material should be closer to the 'low density' end of the continuum – achieving a

high enough density that it will support its own weight, resist damage from things knocking against it, and carry the render and plaster finishes, while remaining as low density as possible in order to maximize insulation and minimize material cost.

The density of the finished material is partly defined by the amount of binder in the mix, which varies depending on the application, for example floors, walls or roof insulation, with different ratios specified to achieve different densities for specific applications. Another factor that can affect the density is the amount of water: as described in the previous chapter, excess moisture in the mix can increase the density of the finished cast material.

Assuming the correct ratio of binder to hemp is used, and the mixing technique is good and hasn't allowed excess water to enter the mix, then the only thing that can affect the density of the finished material is placing technique.

Apart from density, the other important quality of the finished material is the open structure of the surface of the hempcrete wall. The flat surface formed by the shuttering board is covered with small pockets or openings formed by the individual interlocking pieces of hemp shiv in the mix. This open texture is useful for three reasons: it provides the ideal key for lime renders and plasters, it improves the breathability of the finished wall, and it speeds up the drying of the freshly cast hempcrete by connecting the outside air with the many small air-filled channels inside the cast material.

The degree to which the surface is open (lots of relatively large holes) or closed (fewer and smaller holes) depends on how much the hempcrete that lies against the shuttering is compacted during the placement process. More compaction at the edges means a less open structure.

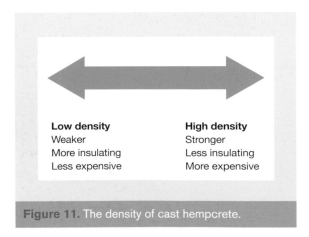

Low density
Weaker
More insulating
Less expensive

High density
Stronger
Less insulating
More expensive

Figure 11. The density of cast hempcrete.

The significance of this is that there is a balance to be struck: a very open surface, while desirable for quicker drying times and better breathability of the finished wall, will also swallow up a lot of basecoat when you come to apply the finishes. This can significantly increase the material and labour costs in the application of finishes.

In general, it is probably best to err on the side of the surface being more open than closed, since the benefits of shorter drying times and good breathability probably outweigh the extra expense in the finishes. However, it is important to keep this balance in mind when perfecting your technique, and to demonstrate to all members of the placing team what level of compaction is required at the edges, and explain the thinking behind it.

Occasionally a more closed finish is desirable, for example where you know in advance that the hempcrete is to be left bare internally as a feature wall. In this case a more closed-up surface is required to consolidate the surface of the wall, as this will not be achieved with plaster. In such

When consolidating the surface of a hempcrete wall, look for a good balance between an open and closed structure. This close-up shows a good consistent surface: not too open and not too closed.

cases the longer drying time resulting from the closed surface is not a problem, as there is no pressure for the wall to dry quickly so that finishes can be applied.

To some extent, however carefully we describe it here, finding the correct placing technique necessary to achieve the correct density throughout the wall and the desired surface texture is a process of trial and error, and first-hand practical experience is essential.

When building with hempcrete for the first time, or with materials you have not used before, it is normal practice to construct a test panel to try out your placing technique and to review its success (see page 213). In this way any problems can be identified, and adjustments to technique made, before starting on the build proper.

Method

The placing of hempcrete is not an exact science, owing to the non-standardized nature of the plant-based aggregate and its low-tech method of application. Although it is by no means a difficult task, the success of the finished cast material depends on the level of experience of those placing the hempcrete.

Always keep in mind the relationship between compaction of the placed material and eventual density and insulation value. There is a risk, with inexperienced or over-enthusiastic workers, of over-tamping the hempcrete, resulting in a higher density. Not only does this mean a lower insulation value of the finished material but it also means increased material costs, as more hempcrete than necessary has been crammed into the void.

The challenge, especially with a large team, is to keep the density consistent across the build. A

Hempcrete is tipped into the shuttering in layers of 100-150mm.

Hempcrete when first tipped into the void.

The loose hempcrete is spread around evenly, and patted down.

finished wall that includes sections of differing densities is not as stable as a wall with a consistent density throughout. For this reason it is recommended that one person, or a very few experienced people, retain responsibility for training new workers and monitoring the consistency of the placing as the build progresses.

Basic method

While placing hempcrete is not as hazardous as mixing it, it is important to wear the appropriate personal protective equipment (PPE) – see Chapter 9. Ensure that, as a minimum, everyone is wearing long sleeves with a pair of chemical-resistant PVC gloves and a pair of latex gloves underneath to protect from any stray particles of lime-covered hemp finding their way inside the PVC gloves. Especially when placing for several days, encourage people to use barrier cream and moisturizer to avoid lime drying their skin out and then working its way into the cracks.

Before beginning to place hempcrete, check around to ensure that everything that needs to be fixed before the placing starts has been completed, for example natural drip details at the plinth (see Figure 14(b), page 246).

The basic method for placing hempcrete is as follows:

- Tip the contents of a tub of hempcrete into the shuttered void so that it forms a layer of loose hempcrete around 100-150mm deep but no more. Any deeper than this and it will be difficult to compact evenly and consistently, leading to loose or dense areas in the finished material.
- With a gloved hand, spread the mix out evenly to make sure it fills the whole void to the same level.
- Pay particular attention to any hard-to-reach areas, such as small corners around doors and windows and around horizontal frame timbers (see page 213).
- Consolidate areas around the frame timbers and against the shuttering using the tips of your gloved fingers, always bearing in mind the desired finish on the surface of the wall (see page 209). If using permanent shuttering, there is no need to consolidate the hempcrete next to these boards.
- Spread it evenly once more to ensure that the compaction is consistent, and check that the above areas are still well consolidated.
- For hard-to-reach areas you can use a stick, e.g. a length of roofing batten, as an extension of your hand, using the same method as with your hand. Apart from this instance, avoid using 'tamping sticks' generally, as this tends to encourage over-compaction.
- At external corners, and at window reveals, a little extra compaction of the hempcrete as you place it is a good idea, to give these

vulnerable areas a little more strength in the set material, and thus maximum resistance to knocks and bumps.

- For certain applications – usually in solid-wall insulation or historic work – where a thinner section of cast material is being placed (e.g. 150-200mm), it may need to be tamped more firmly than usual to increase the strength of the material, but this should still be kept to a minimum. If you are casting that thin then you don't really have much leeway for a reduction in thermal perfor-mance. If in doubt, experiment by casting a test panel at a low density, letting it dry for a week or two, and testing the strength of it.
- When placing hempcrete over any opening in the wall, the section immediately above the opening up to at least 300mm above it, and extending at least 300mm either side, needs casting in one go, so that a single cast-in-situ 'lintel' of hempcrete is formed. Where the opening is not formed by permanent shutter-ing then some sort of reinforcement needs casting into the hempcrete to support it while it dries, e.g. battens spanning the opening (see Figure 37, page 339).
- When you reach the top of a shuttering section where no hempcrete will be placed directly on top of it, e.g. a windowsill, over-fill this section so that the hempcrete sticks out of the top of the shuttering. This is necessary because the exposed section at the top can dry too quickly and become loose and friable. Over-filling allows the loose hempcrete to be cut off after it has dried (using an old saw or multi-tool), back to the solid stuff at the required level (see photo overleaf, bottom left).
- Ensure that the wet hempcrete mix is placed well before it has started to take its initial set. This is not usually a problem, unless using Prompt Natural Cement, which has quite a short window of workability – though this can be increased a certain amount through

the judicious use of Vicat's Tempo (citric acid) retardant (see page 46). If you use Prompt after it has started to go off, strong bonds are not formed in the cast material, leading to loose mix which can disintegrate when the shuttering is removed. (Thankfully such patches can be remedied if they don't extend too far: see page 214.)

- Although no manufacturer specifically recommends a maximum daily lift in their literature, it is much better as a general principle to move around the whole building to the same level before moving up again. This is so that the placed hempcrete has the maximum possible amount of time to begin setting before extra weight is placed on top of it. Unless your walls are very short, meaning that you would complete each lift very quickly, you are unlikely to have any difficulties in this regard. We have occasionally done lifts of 2 metres in a day without experiencing any problems.
- When a day's work does not take you to the end of a run of shuttering, it is important not to come to a 'sudden' stop. Vertical day joints are to be avoided, because they have the

potential to weaken the cast material. Instead, slope the end of the placed hempcrete down gradually over several metres, so that a gentle slope is formed as the day joint, and build up from this the next day.

A section of shuttering filled with hempcrete.

Hempcrete is over-filled at windows and then cut back with a saw so there is solid hempcrete immediately under the window.

Hempcrete wall showing lines left by shuttering lifts, and a gently sloping day joint.

Bonding between lifts

The act of creating a bond or 'key' between lifts may be a necessary part of placing the hempcrete, or it may not be necessary at all. This depends partly on the design of the wall and partly on the materials being used. The procedure recommended by your material manufacturer or supplier should always be followed.

In the case of a central frame design, additional bonding measures are usually unnecessary, since the frame itself extends through the centre of all the individual lifts of hempcrete, tying them together.

With other frame designs, extra measures to ensure a good bond between lifts might include, for example, casting sections of batten sticking out of the top of a lift centrally at regular intervals, so that when the next lift is cast the batten ties the two together.

Other measures might include (especially with a fast-setting binder) wetting down the top of the previous lift with a spray, or painting it with a slurry of binder and water to control suction (the action by which a very dry surface can pull too much water out of a lime binder), and thus encourage good adhesion between the two layers.

Placing around horizontal frame sections

Where horizontal timbers form part of the frame, extra care needs to be taken to ensure complete and even placement of the hempcrete at a consistent density around and especially *under* these frame members. Because it is so hard to reach into the shuttering and shove the last bits of hempcrete under a horizontal timber, the tendency is for small sections to get missed, or for the hempcrete in these areas to be either too loose or over-compacted.

For this reason, wherever possible, horizontal timbers should be kept out of the frame design, but in certain circumstances, including the frame around windows, timbers for electricity sockets to fix to, and the extra horizontal battens used in an exposed frame design, it is unavoidable.

With an exposed frame design, horizontal battens as 'rails' are fixed to the upright studs at 600mm centres going up the frame. To make things easier, some hempcrete contractors recommend fixing these rails as the build progresses. Since the standard shuttering lift is also 600mm, the shuttering can be filled to the top, and a rail laid along the even flat top of the filled shuttering and fixed, then a new shuttering board is fixed and more hempcrete placed on top of and either side of the batten. This is a great idea, as it allows the easy placement of even density hempcrete all around the batten, and the horizontal batten doesn't get in the way as you are tipping the hempcrete into the shuttered void. However, on a large, busy build this may not be compatible with the way that the hempcrete casting has to fit in around other works. Whole-building runs of shuttering before another lift is started are not always possible in practice, and on a busy site the job of fixing the rail before the next lift of shuttering goes up sometimes gets forgotten. Since these rails provide such a vital function, you might decide it is safer to fix them all to the frame at the outset, before placing any of the hempcrete.

Reviewing the success of hempcrete placing

It is essential to review the success of your team's hempcrete placing, so that mistakes can be picked up early and adjustments made to the placing technique. This needs to be done after building a test panel, at the end of the first day's placing, and then again as necessary throughout the build.

Before it sets hard, hempcrete can easily be shaped using a nail float. Here the sharp corners are rounded off to give softer lines to the building.

cavity allows faster drying of the hempcrete than does a standard render finish.

It is important that finishes are not applied too early, for two reasons:

- First, because the drying out process can be slowed significantly when there is still a very high water content in the wall, there is an increased risk of the hemp shiv rotting, although in reality this is probably a fairly unlikely outcome.
- Second, and much more likely, is the risk of staining on the plaster or render finishes (see Chapter 5, page 74). The reason for this is that tannins present in the hemp shiv can be carried through in water droplets and deposited on the surface of the render as the water in the wall evaporates. These unsightly stains can be painted out when the wall has dried sufficiently, but if the reason for rendering early was to reduce scaffolding costs, you will be disappointed to find that it is even more expensive to have scaffolding reinstalled so the walls can be painted again.

The surface of hempcrete quite quickly forms a dry crust, and there is often scheduling pressure to apply finishes as soon as the wall 'looks' dry. This is especially the case with external renders, stemming from a wish to take down scaffolding at the earliest opportunity to reduce costs.

Although finishes for hempcrete are always breathable coatings, and must remain so in order for the wall to work properly, they do slow down the drying of water out of the wall once applied. This is because the finish covers up the open structure of the bare hempcrete, reducing ventilation to the surface. In the case of plasters and renders, the open structure of the hempcrete surface is closed up by the finish, reducing the overall surface area from which the water can evaporate. A cladding finish with a well-ventilated

Render applied too early: dark patches show where liquid water is drying through the render.

When the surface has dried sufficiently, finishes can be applied.

The way to tell whether the wall is sufficiently dry for the application of finishes is to ensure that there is a consistent dry crust on the surface of the wall, to a depth of around 40mm. This indicates that the wall has dried to the extent that water is now reaching the surface of the wall only as moisture vapour, rather than liquid water droplets. Tannins only dissolve in liquid water; they are not carried by moisture vapour, so once this stage is reached you are safe to apply renders and plasters without the risk of staining.

The time taken for hempcrete walls to dry to this point varies depending on a number of factors: the type of binder used, the thickness of the wall cast, local weather conditions, good mixing technique (limiting the amount of water that goes into the mix) and good drying management once walls have been cast. The 'normal' time before finishes can be applied is around 6-8 weeks, depending on the binder used and of course on the thickness of the wall.

When planning the build schedule, 6-8 weeks after casting should be considered the earliest time at which finishes (especially wet finishes) can be applied to a wall of 300-350mm, *and only if local conditions are favourable and the correct mixing and placing techniques are followed.* Additional factors, such as the time of year and on-site management of drying (see Chapter 19, page 264), will also have a significant effect on drying times. For example, in particularly wet climates or at colder and wetter times of year, drying can take significantly longer. Sustained wet weather brings the double problem of it

CHAPTER SEVENTEEN

Floors, ceilings and roof insulation

The previous two chapters describe the procedure for mixing and placing hempcrete for its 'normal' use: within the walls of a building. However, it is also possible to use cast-in-situ hempcrete in other applications: as insulation in a vapour-permeable or 'breathable' floor, in a ceiling, and as roof insulation. These applications are discussed here, with particular reference to how the methods differ from that for the standard walling application.

Floors

Hempcrete can be used as an insulating, vapour-permeable floor slab. Note, however, that breathable floor slabs are not suitable for areas in which a radon barrier is required in the floor slab.

Conventional floor construction involves the use of synthetic insulation products, such as stiff polyurethane boards, together with concrete, a plastic damp-proof membrane (DPM) and a cement screed to create a non-vapour-permeable floor. Synthetic insulations have high embodied energy, and carry the risk of off-gassing of toxic chemicals from the floor. Additionally, since it is not breathable, this type of floor can cause problems with the management of moisture in the ground immediately below the building. In particular, when retrofitted to old buildings that were originally built with a breathable floor and walls with no DPC, this modern floor construction can cause rising damp problems, as moisture which cannot rise through the floor is instead forced up the walls in more concentrated amounts. Often, the use of cement render externally and gypsum internally exacerbates the problem

Left: Hempcrete floor, walls and ceiling make it easy to achieve airtightness around the whole thermal envelope.

Underfloor heating can be included above the hempcrete slab, in the screed or mortar, depending on the desired floor finish. Again, all finishes for a hempcrete floor must be, and must *remain*, breathable (see next chapter).

Method

The basic steps for creating a breathable hempcrete floor are as follows:

- Excavate the floor area to the necessary depth, add hard core and level this out.
- Use a suitable-sized gravel to level out any significant dips in the hard core surface, to avoid having to fill them at the next stage with the (more expensive) insulating sub-base aggregate.
- Pour in the sub-base layer. This should be free-draining, non-capillary and insulating. It usually consists of loose particles of coated expanded clay aggregate, or recycled glass foam aggregate. This stage is easy work, as the aggregate is very lightweight.
- Ensuring that the appropriate PPE is worn, mix the hempcrete according to the manufacturer's specification for a floor, and place the hempcrete slab. The difficulty here is getting it level. Hempcrete doesn't spread out easily like concrete, as the particles of shiv give it a dry consistency. Draw a level line around the perimeter walls at the required height and use long levels or a laser level to guide you, as you work backwards towards an exit.
- For the hempcrete layer you are aiming for a 'generally flat' surface, not a completely smooth one. Any small bits sticking up are covered in the screed or mortar that is usually applied over it.
- Once the hempcrete has taken its initial set (overnight for most binders, or 30 minutes for Prompt), you can walk on it, but try to limit this as far as possible to light traffic for the first few days.
- Give the hempcrete plenty of time to dry out fully (see Chapter 16, page 215), and follow good drying management throughout this period to speed up the drying process as much as possible. Be particularly careful to not leave anything on the floor.
- Cover the floor temporarily to protect it if other works are being carried out.
- When the hempcrete is dry you can apply your chosen breathable finish (with integral underfloor heating if required). These include, for example, natural stone or clay tiles set in a breathable lime mortar, a limecrete polished screed, a compacted earth floor, loose-fitting timber boards, or an engineered timber floor with a vented air gap underneath (see next chapter).
- If the final finish will be a timber floor, then battens can be cast in the floor, flush with the surface, for the timber floor to fix into (as pictured on page 85). The inclusion of battens makes levelling the hempcrete slab a much simpler process: you can level the battens first, and then fill in around them with hempcrete. Since these battens provide the main fix for the floor above, they need to be wide rather than deep in order to spread the load (although in practice the floorboards will also bear directly on to the hempcrete). In the past we have used 80mm x 25mm garden decking boards to good effect, with the grooved side facing down to help with key against the hempcrete.

Ceilings and roof insulation

When cast around a reinforcing timber frame, hempcrete can be used to form ceilings and provide insulation at roof level. Although not widely used in this way in the UK, these applications are perfectly possible, and all of the binder products currently on the market are suitable for them, according to their manufacturers' technical specifications.

At Hemp-LimeConstruct we have successfully used cast-in-situ hempcrete for both roofs and ceilings, usually in situations where it was important to the client to keep the use of processed, high-embodied-energy materials to a minimum.

Ceilings

Cast hempcrete is suitable for ceilings inside either flat or pitched roofs, or between floors where there is cast-in-situ hempcrete insulation directly above the ceiling. Hempcrete ceilings are never cast independently of hempcrete insulation. Cast-in-situ hempcrete is rarely used for downstairs internal ceilings and insulation between floors, since the extra expense is prohibitive, unless increased acoustic or thermal insulation is required between different parts of the building.

In comparison with conventional ceiling boards, a cast-in-situ hempcrete ceiling has a number of advantages:

- It provides insulation.
- It reduces material costs.
- It reduces the number of different materials used in the building, thereby reducing the complexity of achieving airtightness.
- It has the same environmental advantages of hempcrete in other applications – being made from renewable materials with low embodied energy and relatively low transport costs. In contrast, conventional ceiling boards consist of highly processed materials, often come from abroad, and may include glues and sealants that contain VOCs.

Another advantage of using cast hempcrete in ceilings, as with other applications, is that it is an ideal background to plaster on to with lime or clay plasters, and provides a breathable ceiling which can help regulate humidity in the room. This is especially useful in bedrooms, where the

Erecting shuttering for a hempcrete ceiling.

CHAPTER EIGHTEEN

Finishes for hempcrete

Finishes applied to hempcrete floors, walls and ceilings, as to any other substrate, perform the function of consolidating the surface, protecting it from damage and preventing excessive moisture ingress. In addition, as readers who have not turned straight to this chapter will be aware by now, they must also remain 'breathable' – vapour permeable. In most cases the finish for hempcrete will be a lime-based render, plaster or screed, but there is also a range of other options, including a variety of natural cladding materials for the exterior or, less commonly, the interior.

In choosing exactly which finishes to use, and how to apply them, people without experience in using natural finishes are advised to stick to the advice of the hempcrete manufacturer or supplier when it comes to specifying and applying a finish. For those who are well versed in the use and performance of natural finishes, however, a wider choice exists. Hempcrete provides an ideal background over which to apply many different natural finishes. This chapter explores the peculiarities of hempcrete as a substrate compared with other common backgrounds, and describes how to apply an example finish appropriate to each building element.

The application of plaster and render finishes, especially when using lime and clay, is a skilled trade, and it is beyond the scope of this book to provide instruction in all the techniques necessary to achieve a fine finish with these materials. Nor is it possible here to give in-depth information about all the other possible finishes for hempcrete and how to apply them; instead, we aim to list appropriate finishes and describe any specific adjustments required by the hempcrete substrate. We strongly advise further research in this area for anyone building, specifying or designing with hempcrete.

For those who wish to learn more about a range of natural finishes, including lime and clay, we heartily recommend *Using Natural Finishes: A step-by-step guide* by Adam Weismann and Katy

Left: This oak staircase, made from reclaimed beams, complements the soft lines of the hempcrete walls.

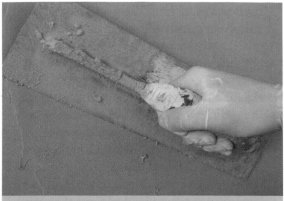

Gradually angle the trowel towards the wall to apply all of the plaster.

The right timing is crucial when rubbing back the plaster to achieve a finish.

Rubbing back lime plaster

There are important reasons for rubbing the plaster back with a float. When the plaster is laid on the wall, the action of the trowel pushes the aggregate into the plaster, leaving a fine, smooth layer of set lime and water, which is less vapour permeable as it contains very little aggregate. The rubbing action removes this thin layer, 'opening up' the surface and exposing the aggregate just below it, leaving a surface with lots of tiny bumps caused by the particles of sand.

In the basecoat, this has the effect of:
- increasing the total surface area for quicker and more consistent drying of the plaster
- providing a better key for the topcoat
- improving vapour permeability between the coats.

Also, in the topcoat, the rubbing action gives an attractive and consistent texture to the plaster.

Depending on the aesthetic effect desired, floats made from different materials and/or different techniques can be used.

basecoat, but using the correct ratio and aggregate for topcoat.
- Wet the wall down with the sprayer until it stops taking water in (see box on page 242).
- Apply the topcoat using the method described as for the basecoat, and rub it back with a fine rubber float instead of a coarse one.
- Once it is rubbed back, go over it with a damp sponge (the big yellow car-cleaning type) in the same circular motion you used with the rubber float. This removes loose aggregate and gives a smoother surface.

When applying a *render* topcoat, a smooth surface is less desirable, since the increased surface area given by a rougher finish allows quicker evaporation of any water that soaks into the render after heavy rain. This can be achieved by floating the surface with a wooden float or a metal grid float, or scraping the edge of a metal trowel across the surface. These techniques will also give a more consistent colour to the surface of the render, as the aggregate is more evenly distributed on the surface.

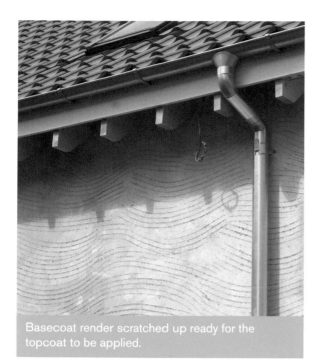

Basecoat render scratched up ready for the topcoat to be applied.

Lime plaster and an oak frame in a new-build hempcrete house.

Wetting down hempcrete to ensure the correct level of suction (see box overleaf) before applying a basecoat plaster.

CHAPTER NINETEEN

Practicalities on a hempcrete build

Having set out the 'nuts and bolts' of hempcrete construction, we now consider the wider practical issues on a hempcrete building site. The main roles are discussed here, together with factors affecting the success of a hempcrete build. Many of the issues raised in this chapter will be especially important to those who are building with hempcrete on a busy commercial site, alongside a range of other contractors. If you are going to be doing a lot of hempcrete work, or your project is a large one, then this chapter is for you. If you are a self-builder with only a one-off project to think about, you won't need to worry about *everything* contained in this chapter, but a broad understanding of the concepts set out here can only improve the efficiency of your build.

Roles on a hempcrete build

As we have seen in previous chapters, there are three distinct stages to a hempcrete build. Often these are actually separated by a period of time, as well as conceptually. The roles at each stage are summarized below.

Stage One: framing

A supervisor is required to take responsibility for the frame being constructed to the structural engineer's specifications, and for the requirements of the hempcrete that is to be cast around it.

A team of joiners either builds the frame on-site from plans, or assembles a frame that has been

Left: A pan mixer for hempcrete running off a telehandler on a large building site.

The person mixing must be kept supplied with binder and hemp.

can be brought up while mixing continues. If a telehandler is needed to move materials and this will stop the mixer, a temporary solution is for the extra team members to move (by hand) enough materials to keep the mixer going until the next scheduled break, when the telehandler (and the whole build team) is used to replenish materials, since the placing work has stopped anyway.

Ferrying team

Scenario: As the day has gone on, the part of the shuttering that was being filled in the morning (10 metres or so from the mixer) has been finished, and work has moved to a part of the building that is 35 metres from where the hempcrete is being mixed.

Problem: The extra time taken to ferry the mix to the shuttering might slow the placing team down,

as they have already placed what they have got, and are waiting for more to arrive. It might also slow the mixing team down, since they are waiting for the tubs to come back so that they can empty the mixer and begin the next batch.

Solution: A couple of people from the placing team, or one from the mixing team and one from the placing team, are reassigned to help the ferrying team temporarily before returning to their duties.

Placing team

Scenario: After finishing a long 'clear' run of wall, the tampers reach a section where there are several windows close together, making it a fiddly business to place the hempcrete in, under and around the more complicated frame structure without leaving unfilled gaps.

Problem: Slowing of the rate of placing may result in a queue forming of tubs full of mix sitting on the scaffolding waiting to be placed. This can slow the mixing team down, as empty tubs don't get sent back with the ferrying team, so the mixer takes longer to empty. In bad weather it can also allow tubs to get rained on, meaning that extra, and very much unwanted, water is introduced into the mix. Likewise, in hot, dry weather, the hempcrete in tubs risks drying out too much before it is placed.

Solution: Either a couple of people from the ferrying team can be assigned to work with the placing team, or the ferrying team, instead of just dropping full tubs and picking up the empties, can begin pouring the mix into the shuttering ahead of the tampers, emptying the tubs to return to the mixer. In the second case, it's important that the placing team retains responsibility for which sections of shuttering have been tamped, so that well-meaning ferrying people don't put a layer of loose mix over another loose layer, which hasn't yet been tamped.

Shuttering team

While the work of mixing and placing hempcrete is under way, don't forget to check on the shuttering team. Although they are working fairly independently of the other teams on-site, their job also determines the speed at which the placing work progresses, since they are removing sections of shuttering from cast hempcrete walls and re-fixing them on the section of the frame where the placing team will arrive next.

As with the other teams, the speed of this work is partly dictated by the different sections of the frame: long, straight runs are likely to go up and come down quite quickly, but any corners or fiddly details, or anything which involves cutting boards to shape (such as inside a roof space or on a gable end) can slow the work down considerably.

The shuttering team usually consists of two people.

As we have seen, the accuracy of the shuttering work is vital to the straightness and plumb of the finished walls, so there is no point in rushing it or assigning unskilled workers to this team, other than for fetching and carrying duties or as 'the other pair of hands' to a skilled worker.

Ideally, one or two of the placing team should have sufficient joinery skills to be able to help out the shuttering team when they are being caught up by the placing team. This might include cutting pieces to length, or removing old shuttering and collecting the screws and spacers, under the direction of the shuttering team.

At times, especially on a busy site, it can be helpful to stagger breaks so that the shuttering team can get a 'head start' on the shuttering before the others come back and get the next mix on.

The balancing act: summary

If the issues outlined in the scenarios described here are spotted by, or communicated to, the supervisor as soon as they appear, then the number of people assigned to each team can be adjusted accordingly to keep the speed of work going at its full potential. This might take a while to sort out if you haven't managed a hempcrete build before, but it soon becomes apparent how the flow of people can be managed to ensure that there isn't too much standing around.

After a while, experienced teams can even start to readjust themselves as necessary, adapting flexibly to the fluctuating demands throughout the day, based on their assessment of what is happening at any particular time and on the skills of each individual.

This concept of switching people between the teams is key to a successful build, particularly on a large job. It makes it even more important, in teams that include people with different levels of skill, and possibly volunteers, that a good level of supervision is maintained by the experienced team members, so that mistakes are avoided and the overall quality of work is kept consistent.

Smaller builds

While the process described on the previous page, of streamlining the placement of hempcrete, is important on larger, commercial builds, self-builders and those on smaller builds will be able to work at a more relaxed pace, with fewer people. A smaller build might involve only two to four people in total; everyone might first be involved in putting up a run of shuttering, then all work together to mix and place the hempcrete.

For the mixing and placing stage, one person can run up a couple of bell mixers at a time, while the other two or three people ferry and place the hempcrete. Another person can act in a hybrid role, helping the mixer by fetching more materials when needed, and also taking buckets to the placers and helping them as required.

The 'feel' of working on a small hempcrete build is very different from that of a large commercial build, with increased flexibility of roles, and a more relaxed atmosphere generally.

Site organization

With a bit of forethought, a hempcrete build can (within the limits of a particular site) be organized in such a way as to make the job easier. The following advice on site organization concentrates particularly on the second stage of the build: mixing and placing the hempcrete. Issues relating to site organization during the timber-frame assembly and the application of finishes should either already be familiar to those with experience of conventional builds or be found in general resources relating to the construction industry.

Storage of materials

The area for the storage of raw materials (hemp shiv and binder) needs to be clean, dry and cool, and close to where deliveries are arriving. It also needs to be accessible by forklift or telehandler, if you have one on-site, or by hand pallet truck if you are relying on the lorry driver. If it isn't accessible to one of these, you are going to be doing a lot of lifting and carrying.

Although far from ideal, pallets of binder and hemp *can* be stored outside for short periods, as long as you keep the binder pallet well covered once the original pallet wrapping has been removed. The hemp is usually supplied in plastic-wrapped bales, so rain is less of an issue, but they never stay 100-per-cent waterproof, especially as time goes on and they get knocked and snagged by people working around them.

A little bit of wetting of the hemp shiv, if it is to be used in the next few days, is not a complete disaster, but look out for any areas where the hemp is starting to turn black: this is a sign of rot starting, and these bits need to be pulled out and discarded rather than introduced into the mix.

Storage of any timber (usually untreated softwood, roofing battens and OSB sheets) should be under cover, and the same goes for any board used for permanent shuttering that is not of external grade. Wood wool board is external grade and can stay out in the rain, although we usually cover it just to be on the safe side.

Mixing area

From storage, the materials will be moving to the mixing area. The choice of mixing area is discussed in more detail in Chapter 15, but, essentially, you need to think about how easily you can transport materials there from the storage area, and how a reasonable quantity of materials can be stored next to the mixer and kept clean and dry. The mixing area will ideally be under cover but well ventilated, and ideally slightly apart from areas where other people are working, because of the binder dust flying about.

Wherever possible, have a clean floor to the mixing area, preferably boarded, so that any dropped hempcrete mix can easily be swept up and returned to the mixer. Needless to say, the mixing area needs to have a good supply of water with sufficiently high pressure, so it doesn't slow the mixing down while you are waiting for tubs, or a tank (or bore hole!), to fill up.

An area for cutting

Although the majority of the joinery work has been undertaken during the framing stage (before the site was cluttered up with people ferrying tubs of hempcrete around), the shuttering team still needs to cut, and re-cut, boards and battens, as the shuttering is taken down

from sections of frame once they are cast and re-fixed to the next section.

As most of this cutting involves 2400mm x 600mm boards, and a table saw is not always practical, a corner of the floor is often turned over to an area for marking out and cutting boards with hand-held circular saws, and battens with a hand saw or a chop saw. Ideally, this job would be done in its own area, slightly away from the messy hempcrete-placing work, in order that expensive power tools are kept as clean as possible and people are not always walking through where you are measuring and cutting. However, in practice a well-defined separate area is rarely available, and on a larger build the cutting area will gradually move around the frame to follow the work, otherwise you end up carrying the heavy boards over long distances.

If one particular area where all the cutting is to be done has to be identified, then it makes sense to put this close to where timber and boards are stored. The distance from the cutting area to where the shuttering needs to be fixed will change throughout the job as the work moves around the frame, but materials are likely to be stored in the same place throughout.

Tools

It goes without saying that you need somewhere secure and dry to store tools if they are to stay on-site. Many people will be happier taking expensive tools away each day, but remember that you will also be storing tubs, buckets, endless pairs of waterproof gloves, goggles, waterproof clothing and boots, first aid kit, many boxes of different fixings, a nail float, chisels, hand saws, levels, squares, tape measures, access equipment …the list is endless! You are unlikely to want to pack that lot up every night to take home, so give some thought to secure lockable storage on-site.

Scaffolding and access

Unless you are working on a single-storey building with a simple roof design, then you are likely to need scaffolding, at least externally. We will not go into detail about this here, as the scaffold required is likely to be similar to that needed for any building project. Suffice to say that all scaffolding should be assembled and taken down by qualified persons, and should remain in place and not be adapted ad hoc during the build. Falls from height on-site are a major contributor to deaths and lasting disabilities caused on building sites, so ensure that you, and everyone else on the team, are familiar with safe working practices with regard to scaffolding and working at height, and insist that these are followed at all times.

The normal procedure is to hire the scaffolding from a firm who comes and puts it up before the work starts and takes it away afterwards. In negotiating this, wherever possible get a 'for as long as it's needed' price, rather than a weekly rate or a price for a fixed time, after which it reverts to a weekly rate. The main reason for this is that there may be an extended time between the walls being cast and external render applied (see Chapter 16, page 216). If the scaffolding has to stay up only because you need to come back later and render the walls, you are paying by the week for something that is not required in the meantime, and striking it and then re-installing it is likely to be even more expensive.

Otherwise, at the time of the scaffolding design, think ahead and consider what exactly is required in terms of the mixing and placement of hempcrete itself. For example, if you will be mixing off the scaffolding, do you need a reinforced platform section where large amounts of binder and hemp can be lifted up with a telehandler and stored on the scaffolding next to the mixer? Or, if you are mixing on the ground and lifting the mix up on to the scaffolding in tubs, is there a way to make this easier, for example with a block and pulley set up on a protruding pole above a removable section of rail? Will you be able to access each lift of shuttering as you move up the wall?

Think about the horizontal poles supporting the platform at each level (i.e. at 90 degrees to the wall).

Think carefully about scaffolding design, e.g. how will you access each lift of shuttering as you move up the wall?

Ensure that horizontal scaffolding poles are not too close to the wall to prevent shuttering being moved.

These need to not only stop short of where the face of the cast hempcrete will be, but also leave enough room between the pole end and the wall to allow large shuttering boards to be removed easily without denting the freshly cast wall. You will also need space to comfortably get a trowel and float in the gap when applying finishes.

Other than scaffolding, other useful access equipment includes stepladders, moveable trestles with height-adjustable platforms, and small folding hop-up platforms (usually about 45cm high). These 'hop-ups' are perhaps the most useful: they can be easily moved around and are often in near-constant use by the teams shuttering and placing the hempcrete, so it's worth having at least a couple on-site.

Somewhere for breaks and somewhere to get clean

Every hempcrete build needs a proper area to wash yourself and your equipment, and to get clean at the end of the day. Mixing and placing hempcrete, especially in poor weather conditions, can be a pretty messy business, and you end up getting covered in the stuff despite your best efforts with protective clothing and gloves. Because of the lime in the binder, the opportunity to clean the worst off your hands and face and out of folds in your clothing whenever you stop for a break is not just a luxury but an important part of preventing lime burns.

You will also want to wash out goggles and thick PVC gloves at the end of the day and put them to dry somewhere overnight, otherwise you will get small pieces of lime-coated hemp inside your gloves to rub against your fingers and cause lime burns, and inside your goggles to fall down into your eyes the next time you look up at the ceiling.

For similar reasons, it is certainly preferable to have a separate area away from the work to sit down for breaks and meals. You soon find that small pieces of lime-coated hemp get spread everywhere around the work area, no matter how well you keep on top of the sweeping. It's asking enough that you have to work with the stuff; you really don't want it getting in your sandwiches or cup of tea as well.

The weather and temperature

As you will be aware by now, the weather is a much more significant factor when building with hempcrete and lime than when using cement-based materials. Nothing is certain except death and taxes, as the saying goes, and if there *are* any other contenders for certainty, then the

UK weather is not going to be very high on that list, so we advise you to consider the points outlined below carefully before starting work.

Protecting your work (and your workers!)

Rain is the major environmental risk factor on a hempcrete build. As we have already seen, it is imperative to keep any excess water introduced into the hempcrete wall to a minimum (while ensuring the presence of *enough* water to activate the binder). Any unnecessary water simply extends your drying time (see Chapter 19, page 264), which is to be avoided at all costs.

For this reason, as discussed in Chapters 15 and 16, site the mixer in such a way as to avoid rain getting into the wet hempcrete, and avoid leaving tubs of hempcrete waiting to be placed where rain (or someone over-shooting with the hose) can add to their water content.

Remember also that all sections of wall where hempcrete is to be placed must be protected from rain during and after placement. If water is allowed to run down through the freshly placed hempcrete before it has set, this not only risks adding excess water but also carries the further risk of the water washing binder out of the wall and compromising its structural integrity.

Last but not least, rain increases the risk of workers getting lime burns, as powder or wet particles that would normally have stayed on the outside of clothing get washed into every nook and cranny by heavy rain. It is almost impossible to stop this, despite your best efforts to protect yourself. However, the risk can be reduced by making sure there is sufficient waterproof clothing on-site: cheap waterproof jackets and over-trousers, and steel toe-capped wellies or rigger boots for a start. Bear in mind that even in light rain, waterproof trousers or knee pads can

Waterproofs or knee pads help to minimize lime burns in wet weather.

prevent burns caused by kneeling on spilt mix on a rain-soaked scaffolding platform while tamping the hempcrete in the shuttering.

In serious rain, put a halt to mixing and placing work, unless your mixer is under cover and you can arrange things such that the sections of wall being cast are filled from inside, so that workers are not getting drenched.

Risk of freezing

The temperature is always a factor when placing hempcrete, for the same reasons as with any lime-based product, including (although to a lesser extent) ordinary Portland cement. What you are trying to avoid is the temperature dropping to below 0°C before the binder has had a chance to react chemically with the water (hydrate) to achieve its initial set.

At around -1°C, some of the water in the mix will begin to freeze, slowing down the speed of the hydration reaction. At around -3°C or -4°C, enough of the water freezes that hydration stops completely and, depending on the extent of hydration and the corresponding strength of the material, the force of the water changing state (ice typically occupies around 9 per cent more space than water) is a risk to the long-term integrity of the setting material.

Clearly then, for hydraulic-lime-based binders to work, a window of time is required when the ambient temperature remains well above freezing point while the binder reacts with the water. For the fast-setting cements this is a matter of hours, but with natural hydraulic lime (NHL) it takes days, depending on the strength of NHL used, with the more hydraulic ones setting faster and harder. Always seek, and follow, the advice of the binder manufacturer or supplier.

As a rule of thumb, work with any lime-based binder (including cement) requires a temperature of 5°C or above. Depending on the type of lime or cement being used, the period of time after application during which that temperature needs to remain constant varies. As an example, when working with NHL 3.5, you really need the minimum temperature to remain above 5°C for 7-10 days, with the risk of weakening due to a sudden temperature drop decreasing with each day that passes, as more hydration occurs.

Although never used as a hempcrete binder in their pure state (i.e. without a pozzolan added to make them more hydraulic), it is worth noting here that air limes (non-hydraulic limes), which set through carbonation in contact with air, set very slowly indeed; sometimes the mortar inside a wall never sets completely. Because of this slow setting time, these limes are even more susceptible to the effects of temperature, and this includes the risk of excessive heat drying out the surface of the mortar or plaster and causing cracking. If you are using these limes – for example, in fat lime

(lime putty) plasters – ensure that you understand exactly how the material will behave in a given situation. At least one person on the team needs to be fully conversant with the preparation, application and after-care of these products, or you are more than likely to run into trouble.

When it comes to hempcrete binders, individual binder product manufacturers specify the minimum temperatures required for working with their products, and you should follow this advice. However, it is advisable to familiarize yourself early on with the temperature parameters required by the product you are using, as this stage of the work will be so much easier and less stressful if you are not racing against inclement weather.

The obvious conclusion is that the best time for casting hempcrete is in the late spring and summer, and this is indeed the preferred option, in terms of both avoiding the risk of freezing and to encourage faster drying times (see below). However, it is possible to cast hempcrete in colder weather, especially if the choice of binder is made carefully.

Prompt, as a natural cement, has a clear advantage here because it is naturally very quick setting and is extremely hydraulic, using up a lot of water from the mix very rapidly. This aggressive take-up of water also reduces drying times – another significant advantage, especially at colder times of the year. All of this allows Prompt to be used successfully in applications where minimum temperatures are not guaranteed over such extended periods as would be required for other binders. However, this does *not* amount to a blanket capability of the product to work in *any* conditions. When working in low temperatures, it is important to ensure that the supplier or manufacturer has signed off on the product's use after being given accurate information about the type of application, including the minimum temperatures expected and over what period, and the thickness of the material you are casting.

As well as thinking about the type of binder to use when working at lower temperatures, take other sensible precautions where practicable. For example, consider casting earlier on in the day to allow plenty of time for the hempcrete to take its initial set before night draws in, bringing falling temperatures. If you are really concerned about a sudden unexpected drop in temperature, then it is sensible to cover freshly placed material with hessian to trap some air against the surface of the wall and keep frost off, although if you find yourself doing this then you really are pushing the limits of what you can do with the material, and it might be less trouble in the long run to take a few days off until the weather warms up again.

Good drying management

The drying time, as discussed in Chapter 16, is one of the most important considerations when working with hempcrete. It is vital that a freshly cast hempcrete wall is given sufficient time to dry out before finishes are applied (see Chapter 16, page 215). In this way the water introduced when mixing is allowed to leave the wall quickly, so that a) there is no risk of the natural materials within (hemp and timber) being exposed to excessive moisture and so being at risk of rotting, and b) the thermal performance of the wall reaches its full potential as quickly as possible.

Since the finish for the wall is also going to be vapour permeable, the hempcrete below would eventually dry out through it, but most finishes will reduce the speed of drying considerably. In an ideal world, a cast hempcrete wall would be left unfinished, at least on one side, for as long as possible until it has had plenty of time to dry, and indeed it is not uncommon for the interior side to be left unfinished while the exterior is completed. But in many cases, and certainly on commercial builds, there is a great deal of pressure to apply finishes as soon as possible, and the slow drying of hempcrete is therefore seen as a negative attribute of the material.

We would hope that the construction industry will adapt to using natural materials, rather than the use of natural materials being adapted for the construction industry. There is no reason, except perhaps extra scaffolding costs or a desire to apply the external render before the onset of winter, why the exterior of a hempcrete building could not be left unrendered while the internal fixings and fittings were done and the use of the building commenced. The external finish could be applied several months later without detriment, as long as the hempcrete walls were not in such an orientation or elevation as to be exposed to driving rain, in which case they would need some temporary protection from the weather.

Currently, however, the construction industry is a long way off from this kind of thinking, and at the present time anything that requires finishes to be applied later than they would be on a conventional build causes suspicion and uneasiness. Drying time will always be an important issue, and is of course dependent on local conditions, but with good site organization and planning there are some obvious things you can do to minimize delays.

The effect of temperature on drying time is easy to predict, since warm air 'holds' more moisture (really meaning that the warmer the air, the more moisture can evaporate into it). Water droplets passing out of the open structure of the unfinished hempcrete wall do so more quickly if the air next to the wall is warmer.

Take appropriate steps, therefore, to maximize temperatures next to the wall. Outside the building, the effect you can have on temperature is limited, though don't underestimate the

In severely exposed locations if casting late in the year, it is advisable to erect a screen around the hempcrete walls to protect them from the worst of the weather while they dry.

importance of clearing away anything that is stopping sunlight from falling on the wall's surface; in particular, make sure that nothing is left leaning up against, or stacked very close to, the wall. Inside, maximizing the temperature might mean sealing up and heating the interior space, if you are really worried that drying is happening too slowly. Avoid using gas-powered space heaters, as these emit a certain amount of water vapour into the air as the gas is burnt. If you have to resort to heating, however, think about what happens as soon as the temperature falls again. As the temperature of the air next to the wall drops, the wall is likely to reabsorb some of the moisture that was released into the air, if it is still present as moisture in the interior space.

For this reason, any sealing up and heating of the inside of the building should be organized carefully to ensure that the moisture that leaves the wall into the warm air is then taken out of the building through appropriate ventilation. On a very basic level, this might sensibly be done by sealing and heating the building overnight when no-one is there, and when work commences the next day, turning off the heaters and opening all doors and windows to encourage maximum airflow through the building. Dehumidifiers can

be used during the night, while the building is sealed and heated, as long as they are used sensibly. All water collected in the dehumidifiers during the night should be taken out of the building to be disposed of, rather than leaving it inside to evaporate back into the internal atmosphere as the humidity drops.

Outside, despite the lower temperature, the greater airflow from wind will have a positive effect on drying times, although of course you are at the mercy of the weather. Don't forget also to think about the orientation of the various elevations, and the effect this will have on the drying time of each part of the building – either positive, through increased exposure to sunlight and wind, or negative, through reduced sunlight or undue exposure to rain. In extreme cases, where a wall is exposed to excessive driven rain, you may need to protect it while it dries, but ensure that this is done with sufficient ventilation to allow moisture to continue evaporating from the wall and to move away once evaporated. In a nutshell, cold, still, damp days will slow the drying process, whereas warm, sunny, windy days will produce the best drying conditions – what works for your laundry will do for hempcrete.

CHAPTER TWENTY

Restoration and retrofit

The original use of hempcrete, and still one of the most important, is in the repair or restoration of heritage buildings. Various terminologies exist to describe old buildings in the UK. In this chapter we use the generic term 'heritage buildings' to encompass both *historic* buildings (those buildings considered to be of historic importance: listed buildings or scheduled monuments, or those in a conservation area) and *vernacular* buildings (traditional buildings constructed before around 1850 by local craftsmen using local materials, and without recourse to formal architecture).

In fact, the vast majority of buildings constructed before the First World War were built using natural materials including lime or earth mortars, plasters and renders, in a breathable solid-wall construction. As discussed in Chapters 3 and 4, the use of vapour-permeable materials enabled the self-regulation of moisture levels within the structure, preventing the build-up of excessive moisture within the building's fabric.

The hygroscopic nature of many of the materials also buffered moisture levels in the indoor air, helping to keep the internal environment healthy for the occupants.

Misguided repair during the twentieth century using hard cement renders and mortars, gypsum plaster and non-vapour-permeable paints and finishes has left many of our heritage buildings in a 'non-working' state. Such repairs often result in high levels of moisture being trapped in the fabric of the building, causing damage to the structure, reducing thermal performance and leading to damp and mould within the living spaces. This has given old buildings an undeserved reputation for being uncomfortable places to live. Thankfully, in recent years, with increased understanding of the issues involved, there has been a resurgence in the use of traditional materials, and the mistakes of the past are being corrected. This is particularly true in the case of historic buildings, and any works carried out on them is strictly controlled by the responsible organization (for example, English Heritage in England).

Left: Hempcrete panels in a Grade II listed Tudor house.

Traditional buildings were always built from natural, breathable materials, such as these clay bricks, oak beams and clay/lime plaster.

However, our architectural heritage in the UK extends further than the 370,000 or so buildings that have a listing, and many people in the UK live and work in buildings built before the First World War. There is growing public awareness of the need to repair, restore and, where appropriate, improve these buildings using natural, breathable materials. Unfortunately, in the context of the current (very necessary) drive to retrofit insulation to older properties, there is again a risk of damage being done to the fabric of older buildings and to the health of their occupants by the use of inappropriate materials.

When asked to upgrade the insulation of older, solid-wall properties, the default response of the mainstream construction industry is to reach for the highest-spec synthetic insulation available, which is then either installed between joists or rafters or glued to the inside face of the external walls using chemical adhesives. Not only do these materials frequently contain harmful chemicals but also they are usually impermeable to moisture vapour, thereby sealing up the breathable wall of the property and causing the sort of problems just described.

In 2012, work undertaken by the Sustainable Traditional Buildings Alliance (STBA), a collaboration of not-for-profit organizations, including English Heritage and the Society for the Protection of Ancient Buildings (SPAB), raised awareness of the issues involved in the retrofitting of insulation to heritage properties and highlighted areas where future research is needed. Their work has led to government-driven retrofit insulation initiatives such as the 'Green Deal' belatedly acknowledging that pre-1919 buildings are vulnerable to harm caused by ill-conceived retrofit measures, and putting measures in place to avoid this happening. In particular, the STBA presents evidence suggesting that buildings of traditional solid-wall construction often achieve much better thermal performance than is expected from modelling, suggesting that the payback from retrofit insulation may be less than anticipated. They also underline the need for the development and use of *appropriate* assessments for traditional buildings, and the importance of a systemic whole-building approach when considering thermal performance, in order to minimize negative unintended consequences of retrofit measures.[1]

As a breathable natural material, hempcrete works in harmony with the existing fabric in older solid-wall properties, while improving the thermal performance of walls that were not usually built from highly insulating materials. Hempcrete was originally developed as a breathable, insulating replacement material for wattle and daub – the mix of earth, straw, dung and occasionally lime woven into a hazel, willow or ash framework to create walls in ancient timber-frame buildings. As well as this specific use, it has a number of other practical applications in the repair of, or retrofitting of insulation to, heritage properties. An overview of these uses is presented later in this chapter.

Hempcrete is the perfect material for repairing and increasing the thermal performance of ancient timber-frame buildings, due to its hygroscopic properties.

Hempcrete also fits against old timber frames much more snugly than board insulation does.

Key concepts in the retrofit of heritage buildings

It is not our intention to set out here comprehensive instructions for the use of hempcrete in heritage properties. Nor would this be possible, or desirable, since in the sensitive repair of or improvements to our architectural heritage every building should be treated on an individual basis, with a bespoke solution designed to provide the maximum insulation while remaining sympathetic to the specific needs of the building – and within the parameters set out by building control or, in the case of listed buildings, by the responsible body as part of the listed building consent. English Heritage has produced a free pamphlet to help owners of listed buildings understand the issues around improving energy efficiency without detriment to the building's character or existing fabric.[2]

About 50 per cent of our work at Hemp-Lime-Construct is on heritage properties, and while we have developed techniques that can often be reproduced in a number of situations, we find

that to some extent each heritage job we undertake becomes a bespoke design-and-build project. The way that the hempcrete is applied in heritage buildings has to be adapted to work with the original building design and materials, as well as to meet the requirements of the client and the conservation officer.

English Heritage are generally supportive of the use of hempcrete within historic buildings, and it is one of the few materials they recommend for the upgrading of insulation in these properties. However, not all conservation officers have had first-hand experience of hempcrete, and some may need convincing before they are prepared to sanction its use. The main issue we come up against is a concern that the hempcrete will be applied in such a way as to alter the building's fabric in a way that cannot be reversed at a later date if necessary. However, this should not be the case, as long as application techniques appropriate to the situation are used, and such objections can usually be overcome with the provision of detailed information to the conservation officer and clear explanations of the proposed work. For example, spray application is not suitable for all aspects of work in historic buildings, owing to the force with which it is projected. When casting solid-wall insulation up against a hard stone wall, the force of projection causes no problems and in fact is an advantage, as it helps the hempcrete adhere to the surface. However, when casting up against the back of surviving plasterwork or wattle and daub, spray-applied hempcrete would adhere too closely to the historically important material, making later reversal of the works impossible without damaging the original fabric. For this reason, in most situations in historic properties, hand-placing is a more appropriate method.

It is recommended that anyone applying hempcrete in heritage buildings should not only be competent in the use of the material but also have specialist skills in, or at least understanding of, the repair of heritage buildings. This should include the repair of timber frames, cob, stonework and heritage roofs, and the use of lime (including fat lime – lime putty – plasters) and, where appropriate, clay mortars and plasters. It is not necessary for the hempcrete contractor to be skilled in every one of these areas, as other specialist contractors can provide these services, but an understanding of all of them is needed in order to be able to take a holistic view of the building and thereby design hempcrete insulation solutions that will complement and enhance the way the existing building fabric works.

The retrofitting of insulation to heritage buildings is a complex undertaking, even once appropriate breathable insulation materials have been identified. This is because the modern conventional wisdom on insulating buildings is to first achieve excellent levels of airtightness by sealing up any potential for draughts and leaks, where heated indoor air can exchange with cold air from outside.

In fact, older buildings were usually *designed* to have air leaks, with the burning of wood, or later coal, needing to draw air in from outside to feed the combustion of the fire or stove. Central masonry chimneys (and often thick stone walls) acted as thermal mass to store the heat produced by the fire, and also as a passive vent, allowing gradual changes of air within the building even when the fire was not lit. While this system was not anywhere near thermally efficient to the

Hempcrete, stone and cob alongside each other in a traditional cottage.

standards we require today, it did provide an effective passive regulation of heat and indoor air quality. More to the point, *it is how the building was designed to work*, so there is no point in sealing up the structure and trying to make it work like a conventional twenty-first-century new-build home.

This demonstrates the importance of having insulation solutions in heritage buildings designed by someone who truly understands how such buildings work. Rather than maximizing airtightness, the appropriate solution will be a balancing act. The idea is to leave the building (or restore it to) working as originally designed, and within this to 'tweak the edges' to get maximum thermal performance out of the system. This might include, for example, installing a wood-burning stove instead of the open fire, adding hempcrete solid-wall insulation (if possible to the *outside* of the wall), fitting natural-fibre loft insulation in the loft, reducing air leaks (but crucially not *all* of them), possibly introducing controlled ventilation, and installing natural insulation between suspended timber floors.

Hempcrete, sometimes alongside other natural-fibre insulations such as hemp-fibre or sheep's-wool quilt insulation, has proved itself capable of bringing solid-walled heritage properties up to a modern standard of insulation if used as part of a sensitive and thoughtful restoration or retrofit of the building. In the case of listed buildings, this is sometimes a little more difficult, because it is necessary to work within the sometimes exacting requirements of the responsible body with regard to maintaining the original appearance of the building inside and out. This can limit, for example, the available thickness within walls or roofs into which hempcrete can be cast. However, under the Building Regulations 2010, listed buildings, buildings in conservation areas and scheduled monuments are exempt from Part L of the Building Regulations (Conservation

of fuel and power), so there is no longer a specific standard to be met. It is also worth noting that, according to English Heritage, special consideration under Part L should be given to "buildings of traditional construction with permeable fabric that both absorbs and readily allows the evaporation of moisture (which can conflict with modern materials and methods)".[3]

The aim for such buildings should be to improve energy efficiency as far as is reasonably practicable without affecting the character of the building or increasing the risk of long-term deterioration of the existing fabric. For a thorough discussion of how to retrofit insulation and renewable energy solutions to improve the sustainability, energy efficiency and comfort of heritage buildings without detriment to their character or to the vapour-permeable building fabric, a good resource is *The Old House Eco Handbook* by Marianne Suhr and Roger Hunt (see Bibliography).

Benefits of hempcrete for heritage buildings

Not only is hempcrete a suitable insulation material for use in a range of situations in older buildings, but in fact there are many situations, especially in listed buildings, where hempcrete is the *only* suitable insulation material. It is worth reiterating that, as with any structural intervention in an old building, hempcrete should be applied only within the context of a holistic review of the thermal performance of the entire building, undertaken by someone with specialist knowledge about the way in which old buildings work. That said, there are significant benefits to the use of hempcrete in heritage properties, which can be summarized as follows:

- Hempcrete works in harmony with, and in a similar way to, original materials in heritage buildings, ensuring that the breathable building fabric is maintained.

- It is more hygroscopic than wood, so it actively 'sucks' moisture away from timber frames and releases it into the air, helping to preserve the timber. This is especially important for frames with externally exposed timbers.

- Heritage buildings are not square and not built to today's standard measurements, and ancient timber frames will usually be quite warped. Wet-mixed hempcrete, being loose-fill, represents a huge saving in labour of cutting insulation boards to different shapes and sizes to fit in between the frame structures. In any case, cutting board materials to fit will never be entirely successful and will leave many gaps in the wall.

- Likewise, being loose-fill, hempcrete fills any voids and gaps when cast as solid-wall insulation against uneven stone walls.

- Finishes suitable for hempcrete are the same as for most heritage buildings, e.g. lime render, weatherboarding, hung tiles, brick or stone.

- Hempcrete, as a cast material, makes it easy to retain the curves and lines of original

Hempcrete can be used to fill small spaces and voids in old buildings, like a natural building equivalent of expanding foam.

building, preventing any loss of original character as insulation is installed.

- Small amounts of hempcrete can be mixed up and used to fill those awkward small gaps and spaces that are always found in old buildings: small voids in walls; gaps behind door frames, at the end of joists, and at the top of the wall under the eaves. It's a breathable, insulating alternative to the menace of expanding foam!

Application method for hempcrete in heritage buildings

The main differences in the application method compared to that described in previous chapters, when it comes to working on heritage buildings, are in the framework and the shuttering. The hempcrete will usually be cast inside the building, next to or around some part of the original fabric, and while the interaction between the two materials can sometimes provide sufficient support, it is more usual for some form of frame to be screwed to the original building to create a key, or a reinforcement structure, for the hempcrete. An example would be a pattern of studs constructed within a timber-frame panel (as pictured opposite), to sit centrally within the hempcrete panel and provide a key against lateral loading from winds (see also Figure 15, page 276). Studs can also be attached to the inside of the frame to take shuttering screws, if the frame is to be left exposed as a decorative feature.

As anyone who lives in one knows only too well, old buildings are not straight, and nor do we want to make them so. Therefore, if applying hempcrete wall insulation externally around a building, for example, this should follow the original lines of the building rather than being cast perfectly square and leaving a structure with pristine

Reinforcing structure within a timber-frame panel.

The same wall with hempcrete cast and shaped.

new-looking straight walls. This is part of the skill of applying hempcrete sensitively in heritage buildings, and inevitably it makes the construction of shuttering a much more complicated and time-consuming business. A bespoke design will usually be required, with careful thought about how the shuttering will attach to the existing building fabric, and be removed, without causing any damage. Constructing shuttering on heritage buildings usually requires a lot of measuring and cutting of shuttering boards. Boards are likely to be smaller than on a new build, and are often needed in weird and wonderful shapes to fit around the existing structures.

Particular attention needs to be paid to how hempcrete walling will interact with the roof at the eaves, and with the bottom of the wall above

ground level. Heritage properties are unlikely to have a damp-proof course, so a robust detail must be designed to ensure that it is not vulnerable to moisture ingress at the base of the wall.

The method of actually mixing and placing the hempcrete in heritage properties will be broadly similar to the method in a new build, although on heritage jobs the total amount needed is often smaller, and the work much more fiddly, than for a new build of the same size. For this reason it sometimes makes more sense to use a couple of bell concrete mixers instead of a large forced-action pan mixer, especially in the case of smaller timber-frame infill panels.

The choice of hempcrete binder used should be considered carefully. Often the limits of what is possible within the context of a heritage building necessitate designs that are inherently more vulnerable to air leakage along the edges of the cast material, so it is best to choose a binder that has very little tendency to shrink during setting (see Chapter 21, page 296), or account for this in the detailing by using hemp-fibre insulation (see Figure 15, page 276, and Chapter 21, page 298). It is advisable to cast some test areas and allow them to dry completely in order to gauge the success of your design (paying special attention to any shrinkage of the hempcrete, and the effect of this on airtightness) before starting on the project proper.

The range of finishes available for hempcrete are in perfect keeping with older properties, since they involve natural, breathable materials. Often, especially for listed buildings or buildings in a conservation area, the external appearance of the building is very important. Whatever is required – whether lime renders, hung tiles or slates, weatherboarding or stone or brick cladding – these can all be applied to the hempcrete walls and constructed from carefully sourced materials to match those originally used in the building.

As discussed in Chapter 2, various types of lime–hemp plaster exist, and the main use of these is within heritage properties. They add some insulation through the inclusion of the hemp fibre in the mix, and, while the insulation value is not as much as with the same thickness of hempcrete, they may be the best option for interior application where space is at a premium. They can be applied in thicknesses of 25-50mm, and it would be difficult and somewhat pointless (from a cost–benefit perspective) to cast hempcrete at such a thickness, as it would require such compaction to achieve structural integrity that the insulation value would be dramatically reduced. Lime–hemp plasters are also very useful for daubing out uneven old walls before they are plastered.

When using hempcrete in a heritage building, always keep in mind the main principles:
- Add as much insulation as possible in a way that does not detract from the character of the building or disrupt the way in which the original fabric was intended to work.
- Apply the hempcrete in such a way that it is reinforced and strong enough to take any loads placed on it.
- Hempcrete should work alongside the other natural materials in heritage buildings. However, it should not interact with them so closely, or in such a way, that the hempcrete could not be removed if required without damaging the original fabric.
- Bespoke solutions are likely to be needed to meet the requirements of the particular building in question.

Hempcrete infill panels to an existing timber frame

Infill panels in a heritage building tend to be much narrower than new-build hempcrete walls, so special care should be taken to create as low density a hempcrete as can meet the requirements for strength and integrity of the material, as the priority should be to create as much insulation as possible within the available space.

The following is our 'standard' method for constructing hempcrete infill panels to an existing frame. This method is adapted constantly to the particular requirements of the building we are working on, and, as already noted, it is more important to achieve the principles outlined above within the constraints of the building and planning conditions than to stick rigidly to the method described here.

Consider carefully any panels that contain original and/or historically important material, such as wattle and daub. If this is intact, the priority should be to preserve it, but sometimes even damaged panels with quite extensive loss of the original material can be preserved, with hempcrete added to replace the missing wattle and daub. The viability of preserving partially damaged panels should be assessed on a panel-by-panel basis (in the case of listed buildings this should be done in close liaison with the conservation officer). The method set out below is likely to vary slightly from building to building.

- Wear appropriate PPE when mixing and placing hempcrete, and ensure safe working practices are followed (see Chapter 9).

- Clean out all material from previous 'repairs': usually this consists of metal lath and cement, but we have also come across attempts at insulation panels using a softwood structure supporting polystyrene boards.

- Construct a row of reinforcing vertical studs centrally within the frame at intervals, specified by a suitably qualified person, sufficient to provide good lateral resistance to the hempcrete panel. These studs can be sawn battens, or coppiced roundwood (usually hazel or ash). They are fixed top and bottom into hardwood battens running left to right centrally within the frame. Their section size should be sufficient to provide the necessary reinforcing strength while allowing them to sit centrally within the frame with a good covering of hempcrete over them.

- If screws to hold temporary shuttering cannot be fixed directly into the outside (visible) faces of the timber frame for aesthetic reasons, separate hardwood battens can be fixed centrally down the inside of the frame to provide a fixing for shuttering screws, which can also be fixed into the vertical studs if necessary (see Figure 15(a) overleaf).

- If desired, a piece of natural-fibre quilt insulation can be fixed between the battens and the inside of the frame to improve airtightness at the edges of each hempcrete panel.

- All permanent fixings should be stainless steel or (in the case of nails) hot-dip galvanized.

- Board out the whole of one side of the panel in advance and then gradually build up shuttering across the other side, filling with hempcrete carefully to ensure all areas within the panel are filled evenly.

- Because the infill panels are generally narrower than for new-build walls, smaller lifts (200-400mm, depending on the frame) are usually required to ensure easy access when placing the hempcrete.

- If appropriate to the character of the building, the shuttering can be spaced out from the frame slightly using battens as packing, and the hempcrete shaped back afterwards to form a soft 'pillowing out' of the infill either side of the frame (see below). This provides an aesthetically pleasing effect while adding a little more insulation than is allowed by the width of the frame alone.

- When the top of the panel is reached, the shuttering board is moved up in smaller and smaller lifts (overlapping the hempcrete below), until a small space under the top frame timber is left. This is filled by pushing hempcrete in from the side, and it is best to do this with freshly mixed hempcrete straight from the mixer, as you are relying on the lateral adhesion of the most recent section of hempcrete to the previous one in order to keep it in the panel.

- When the shuttering has been removed, wait for the hempcrete to set sufficiently for any shaping work to be done using a nail float or multi-tool. Even if the panels are not being 'pillowed out', it is normal to cut the edges of the panel in behind the timber in this way, to about 25-30mm (see Figure 15(b)), so that when finishes are applied they are closed in thoroughly to the timber, with a 'shadow line' effect (see Figure 15(c)).

- When the hempcrete has dried sufficiently, apply finishes as normal. Note that because the panels are normally thinner than in a typical new-build hempcrete wall, the drying period can be shorter.

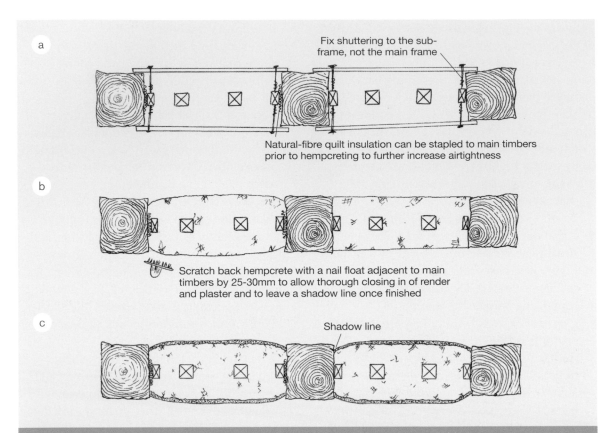

a

Fix shuttering to the sub-frame, not the main frame

Natural-fibre quilt insulation can be stapled to main timbers prior to hempcreting to further increase airtightness

b

Scratch back hempcrete with a nail float adjacent to main timbers by 25-30mm to allow thorough closing in of render and plaster and to leave a shadow line once finished

c

Shadow line

Figure 15. Hempcrete infill to a timber frame with the frame visible on both sides. (a) Shuttering fixed prior to placing. (b) Shuttering removed and hempcrete shaped back adjacent to main timbers. (c) Finishes applied, closed in to the main timbers.

Casting hempcrete as timber-frame infill panels.

Completed hempcrete infill panels before plastering.

Main uses of hempcrete in heritage properties

Of the various uses of hempcrete in heritage buildings, the most common are as follows.

Replacement of wattle and daub in timber-frame buildings

If wattle and daub still survives in a building, it has usually been (and should be) kept for its historic importance. Frequently, however, the original wattle-and-daub panels were replaced years ago, often having lasted many hundreds of years, and then failed after 'repairs' were carried out using a non-vapour-permeable render such as cement. In previous centuries, replacement infill was usually made from bricks laid in lime mortar, which, while not very insulating, was at least vapour permeable – until rendered with cement in the 1960s! This historic brickwork, while not original, is a very attractive addition to the building's character, and one that is often seen has having historic importance in its own right. However, in many cases the weight of the brickwork is too much for the ancient timber frame and can actually cause structural damage to the building. In such cases there is an argument for replacing historic brickwork (where it has to be removed anyway to repair the frame) with an appropriate insulating material such as hempcrete.

If the replacement of original wattle and daub was done in the second half of the twentieth century, it often comprised nothing more than a cement render over 'metal lath' (a kind of steel mesh) on both sides of the wall, with nothing in the middle of the panel. If cement render has been used around a timber-frame construction, this is inherently damaging to the building, as it will hold moisture against the timbers, and should be removed as soon as possible.

Hempcrete acts as a modern insulating version of wattle and daub, and stays very faithful to the original wall build-up, since it contains a fibre (hemp instead of straw), a binder (lime instead of earth or earth–lime) and a supporting wooden framework inside it. At Hemp-Lime-Construct we now normally use local coppiced roundwood for the supporting framework, although in a different configuration from the original wattle. The method of application is as described on page 274.

In the case of timber-frame panels where original wattle-and-daub materials are only partially degraded, it is possible to apply hempcrete in quite a thin layer (50-70mm) by daubing it over a reinforcing structure of split hazel poles, nailed or screwed to the frame timbers. This doesn't add a huge amount of insulation, but very few materials are suitable for this purpose, and hempcrete would add more insulation than the equivalent thickness of lime–hemp plaster, for example, while filling any awkward voids or tricky spaces.

Because wattle-and-daub panels tend to degrade on the exterior face (where they have been covered in cement, which causes more damage than the original exposure), the hempcrete is

Daubed hempcrete over a wattle-and-daub panel, using split hazel poles to provide a key.

most often daubed on to this face, which gives the historic fabric a better level of protection from the elements than a lime render alone, while preserving the character of the building (as shown in the photo on the previous page).

Insulating infill to lath-and-plastered walls in timber-frame buildings

The use of hempcrete cast up against the back of lath and plaster allows the retention of these original materials, usually internally to act as a 'permanent shuttering'. The hempcrete on the external side of the wall can be finished to replicate the original walls, for example with weatherboarding, hung tile or slate cladding or a lime render.

External temporary shuttering is cut to follow the shape of the ancient frame, which over hundreds of years will have twisted and warped to give the building a unique character. If the lath and plaster is fixed directly to one side of the timber frame, and you are wrapping hemp-crete around the frame on the other side, remember to fix horizontal battens to the outside of the frame to provide a key for the hempcrete, as you would in an exposed frame design with internal shuttering.

Solid-wall insulation on masonry buildings

Hempcrete can be cast straight against masonry walls to add insulation. This can be done either internally or externally, depending on the requirements of the situation.

If the wall has been pointed or rendered with cement, or plastered with gypsum, these should be removed (in the case of cement) or hacked off (in the case of gypsum plasters), with pointing replaced with a lime mortar appropriate to the type and condition of the masonry. If pointing is lime and in solid condition, it can remain. It is important to give the wall sufficient time to dry out fully before casting hempcrete.

Clay plasters used in the repair of a historic timber-frame property.

The same repair from the back, before hempcrete is cast around the frame from outside.

The same wall with hempcrete in place.

Hempcrete cast as solid-wall insulation internally fits neatly into the uneven surface of the stone wall.

A completed wall of hempcrete solid-wall internal insulation.

Hempcrete is cast against the face of the wall, either by spraying or by hand-placing using shuttering. Aim to cast as much hempcrete as possible within the situation (and when it comes to calculating the U-value achieved overall, remember that the thermal resistance of the original masonry wall should also be factored in). Depending on the substrate and the thickness being cast, the hempcrete is usually tied in to the original wall, with bolts or screws fixed through, or cast into, the hempcrete. A reinforcing timber frame may be required in some instances, again depending on the thickness of hempcrete cast.

Hempcrete solid-wall insulation is especially useful where walls are of rubble stone or uneven block construction, with irregular stones which create an uneven wall. The hempcrete, being 'loose-fill' in its freshly mixed state, leaves no large voids between the insulation and the face of the stone. In contrast, a board-type insulation on an uneven wall carries the risk of voids being created within the wall, which are then prone to interstitial condensation as vapour passes through.

As it has a good amount of thermal mass, hempcrete complements the original solid walls perfectly, working in the same way but adding insulation. In terms of thermal performance, it is better to cast such insulation externally, as the thermal mass of the original wall is wrapped inside the hempcrete insulation. This also minimizes disruption inside the property, and doesn't subtract any of the living space inside the building. However, casting externally is not suitable for all situations: buildings where the original façade is of historic or aesthetic importance, for example, are usually best insulated internally. Also, casting externally usually involves some small extension of the roof overhang, in order that it covers the top of the hempcrete.

Focus on self-build 2: Hemp Lime House

Leah Wild never intended to build a hempcrete house. She was ready to construct a steel-framed, earth-sheltered home, and had had the designs drawn up, got planning permission, and bought many of the materials. She was on the point of buying the plot of land on which she would build it, when the vendor pulled out at the last minute – leaving Leah, who had already sold her house, with three children and nowhere to live.

Hurriedly looking around at what was for sale locally, Leah found a 1920s railway worker's house: a single-storey chalet bungalow of timber-frame construction, with 5mm asbestos boards internally and externally. "There was nothing at all except the frame timbers sandwiched in between the asbestos sheeting. The suspended wooden floor was rotten, as was most of the frame itself," says Leah. "Absolutely every surface inside the house was covered in asbestos."

Always keen to build using natural materials, Leah was introduced to the idea of hempcrete by a friend. After discussion with William Stanwix, she decided that hempcrete was the perfect way of dealing with the building. "The hempcrete would be cast around the frame, providing it with racking strength and excellent thermal performance. Because it's a relatively lightweight material, it matched what was there in the original structure, so there was no need for expensive alterations to the foundations. The hempcrete was easy to work with, and it looks and feels beautiful. On top of that, the house in its original

state was not considered to be a mortgageable property, but once we'd improved the overall strength of the timber frame and put the hempcrete in, upgrading the thermal performance, I was able to get a mortgage on what we had built."

Because mortgage companies were wary about lending on it, the house was priced cheaply considering its size and its location on a pretty Cotswold hillside just outside the town of Stroud. Leah clearly had the vision necessary to take on this dilapidated, freezing, toxic shell and turn it into the comfortable, healthy, thermally efficient house it is today. She says: "We stripped out all the asbestos, repaired the frame and cast 240mm hempcrete around it." The frame was positioned on the internal surface of the new walls, and wood wool board fixed to it as an internal permanent shuttering. Hempcrete was then cast between the vertical studs and around the outside of the frame. Leah's planning permission also allowed her to build two small extensions, to make the T-plan house into a rectangle. Since these were built in the winter, when using hempcrete would have been difficult, she used hemp-fibre quilt insulation in the walls instead. Finding herself unable to stop, Leah went on to convert an old shed in the garden into an 'annexe', where her teenage daughter now lives, using natural materials in that conversion too.

The hempcrete in Leah's house is finished in a mixture of home-made lime–hemp plaster and lime plaster in different places around the building, in the spirit of trying different things to see how they performed. A locally grown larch

1. Hemp Lime House sits up on a hillside nestled in a lush garden.
2. Leah has proven that a stylish interior can be created from second-hand and reclaimed items.
3. Local larch cladding was used on the gables.
4. Natural plasters and breathable paints create a soft feel throughout.
5. Although small, the house still manages to feel spacious and full of light.

cladding was used on the gables. Most of the plaster hasn't been painted, as "it was a lovely finish without anything being added to it". Where it was painted, a lime paint was used. The bathroom was finished in gypsum plaster, the surface brushed to give it a rough look and three coats of wax applied to simulate a Moroccan tadelakt look.

Leah designed the house "on the back of an envelope", having blown her architect budget on the house that never was, and freely admits she was broke and "had to beg, borrow and blag as much money as I could just to get the building to a stage where I could raise a mortgage on it". The ethos of Leah's build is reusing and recycling: window frames made from pitch pine beams reclaimed from an old church; £6,000-worth of low-emissivity argon-filled glass from eBay for £400; floor tiles, wall tiles and kitchen all surplus stock from eBay; and she has used natural, local and responsibly sourced materials wherever possible. This low-tech, DIY attitude hasn't prevented Leah from creating a beautiful and comfortable home with its own unique style.

There were no issues with Planning, who it seems were sympathetic to Leah's style of building. All the work was carried out under permitted development, and building control approval was sought retrospectively, as "the build just moved on so quickly". It took Leah and her friends just eight weeks (seven days a week) to complete the build, from starting the demolition of the old building to moving in. Leah used a lot of volunteer labour, and friends with relevant skills helped her out when needed. The main people involved in the build were Leah's friend Andy, her son Liam, and his friends Dan and Luke, with plans and elevations drawings provided by Leah's friend Tracey, in order to illustrate to the local authority the proposed development of the building.

The disadvantage of working with hempcrete, as far as Leah is concerned, is that "you are treading an unknown and unpredictable path, and when you phone hempcrete 'experts' they all give you different advice, or only want to give you advice at all if you are using their own products". Despite this, however, and despite the experimental nature of the build and materials used, Leah is happy that overall there have been very few problems, and where these have occurred they have mostly been due to inexperienced builders with too much youthful optimism about, for example, the number of fixings you need to hold a wood wool board up. A few slight problems with the design showed up a few months down the line: for example, a wall with spots of mould appearing on the render caused by water wicking up into, and running down, the render (as there was insufficient roof overhang – with no rainwater goods – and the render had been continued down over the plinth nearly to the ground). Leah says, "We corrected this by chopping out the render and the hempcrete at the bottom, building up the height of the plinth slightly, and extending the roof overhang, and we've had no problems since."

Leah is impressed with how thermally efficient her house is. "It's boiling hot out there today and it's lovely and cool in here. In the winter when you heat the place up and turn the heating off it stays warm for hours and hours, because the hempcrete acts as a heat store." Leah has a wood burner and a gas boiler with cast-iron radiators: "I recycled the old gas boiler that was already in the house, which in hindsight was a mistake – I think if I did it again I'd go for underfloor heating."

Asked about the reaction of friends, Leah says, "Everyone who comes in says it feels amazing – we get really positive responses from everyone who visits . . . it's got a really nice feeling to it – it never feels stuffy, or claustrophobic, or sweaty, or over-heated, or too cold, or . . . it just feels like a living, breathing building . . . you can sort of tell that it's been built with natural materials – it's just got that feel to it."

1. The open-plan living space.
2. The original building. *Image: Leah Wild*
3. The 'annexe'.
4. Leah brushed and waxed ordinary gypsum plaster in the bathroom to simulate a tadelakt finish.
5. Leah in her hempcrete home.

PART THREE

Designing
a hempcrete
building

CHAPTER TWENTY-ONE
Design fundamentals 289

CHAPTER TWENTY-TWO
Indicative detailing 309

Design fundamentals

There are currently no industry standard details for hempcrete in the UK. Each supplier or manufacturer has, to some extent, set out their own standard details for very basic hempcrete walls, floors and roof insulation and currently there is a lack of clear cross-industry standards. In part, this is due to the fact that building with hempcrete is a fledgling industry and new details, methods of construction and products are being developed all the time. Likewise, many of the materials and products that work with and complement hempcrete are also relatively new, for example recycled glass foam blocks, natural wood-fibre boards and hemp-fibre insulation; and this constant innovation, while very welcome in many respects, contributes to the problem of standardization.

Another contributing factor is that until quite recently the hempcrete market in the UK was effectively monopolized by one company supplying both hemp shiv and binder. The amount of hemp building which happened was not sufficient to warrant a more competitive marketplace or widely agreed standards for construction. This has resulted in testing of, and detailing for, proprietary products rather than hempcrete as a generic material. In recent years the market has begun to open up slightly, with more choices of binder and sources of shiv coming on to the market. It is hoped that this situation will create an atmosphere in which wider industry cooperation and agreement on standard detailing will emerge.

A cross-industry acceptance for hempcrete as a material would allow standard details and construction practices to be developed for the most common types of buildings, and for materials used alongside hempcrete. This could only be beneficial to the industry as a whole since it would simplify the building control process, and an accepted construction methodology would allow those building with hempcrete for the first time to use the material with greater ease. It would remove a lot of the stumbling

Left: A slightly tapered window reveal allows more light into the room and is an interesting alternative to a square reveal.

suitable as structural timbers, but could be used for non-structural applications such as supports for fixings or at the joins of permanent shuttering board (see Figure 20).

Topography and climate of the site

There has been a tendency over recent decades, since the development of 'moisture-impermeable' building materials, to virtually ignore the siting and orientation of buildings during the design process. This is probably also due to the increased psychological detachment of many people in the Western world from the weather in their immediate surroundings. Nowadays it is common to spend little time outdoors and to go from centrally heated home to heated-and-air-conditioned car to heated office and back again, without even small and obvious changes in behaviour such as wearing a coat in colder weather. The most reaction many people ever have to the weather is to move the dial on the heating control.

Even if moisture ingress is not (theoretically at least) an issue in modern conventional masonry buildings, the lack of thought about the prevailing weather on-site during the design of new builds leads to high levels of inefficiency in power use as a result of poorly sited buildings. All of us will be able to think of times when we have sat in a home or workplace that was over-exposed to driving wind in the winter, making it too cold, or to direct sunlight in the summer, causing overheating. As well as incurring increased insulation costs, such buildings often need to be fitted with higher-powered air-conditioning or heating systems, which have a direct impact on the running costs and environmental impact of the building. In the last 60 years or so, such inefficiency was not a critical factor in design, but as we move forward into a world where energy resources are increasingly scarce

and costly, such inefficiencies need to be designed out of all new buildings.

Furthermore, it is important to remember the vulnerability of all materials, including hempcrete, to excessive moisture ingress (as discussed on page 293), and the tendency of all walls to exhibit reduced thermal performance when moisture content gets too high. With these factors in mind, it is clear that to use any natural building material to best effect, thought needs to be given to how the topography and climate of the proposed site is likely to affect the performance of the materials. Simple changes to the siting and orientation of the building, the landscaping of the site, or the detailing of the building itself at the design stage can mitigate less-than-ideal conditions. With a little care it is possible to create a building that truly maximizes the potential of the natural materials, with potential problems designed out.

Much has been written elsewhere on this topic, and it is not our intention to replicate it here, but a few of the key issues, as applied to hempcrete, are outlined below as examples of things the designer should be keeping in mind.

Site-specific solutions

There is unlikely to be a 'one size fits all' solution when detailing hempcrete. First, consider the conditions and topography of the specific site and outline the challenges they bring through differing levels of exposure to prevailing winds, rain and sun. Consider the position of the sun at different times of the year – high in the sky in summer; low in winter. Is it possible to design in, through the siting and orientation of the building or through features such as windows and skylights, shade for the interior from the high, hot summer sun while making full use of the low winter sun by collecting the heat it offers into the building? Consider the likely orientation and force of the prevailing wind and rain. If these are of a direction or degree likely to have a detrimental effect on the breathable hempcrete walls, can this problem be resolved through the siting, orientation or detailing of the building?

Using natural features on the site

Wherever possible, work 'passively' to make best use of the natural features of the site that may provide shelter, rather than rushing to the solution of costly and energy-consuming landscaping. Consider the site topography, including slopes, natural banks and existing mature trees, and site the building accordingly to make the best use of the shelter or shading they provide. If this shelter is not thought to be sufficient, then add to and enhance natural features rather than re-landscaping the site. On a completely level site with no mature trees, design the landscaping and planting in conjunction with the siting and orientation of the building to provide maximum shelter and shading as appropriate. Wherever possible, design in the reuse of earth removed during ground works, which will reduce the cost of both landscaping and 'waste' removal from site.

Learning from the past

Take some time to look around the area in which the building plot lies. Any vernacular building (by which we mean any traditional house, farm building, village hall or chapel built before about 1850) was constructed from local, natural materials and was built with a solid breathable wall (see Chapter 10, page 118). In addition, these buildings were built with the full benefit of hundreds of years of local knowledge, handed down through generations, about how to make best use of natural materials within the context of the local topography and climate. It's probable that the standard of workmanship in these old buildings varied dramatically, but those that you

can see around you are a self-selecting best of the bunch, in that they are still standing.

These vernacular buildings have lessons to offer the designer working locally with natural materials. What decisions were made about the siting and orientation of buildings in the immediate vicinity of the site? Look for any obvious details that were designed to mitigate exposure to the weather, such as the west-facing wall on all buildings on one side of a hill being clad in hung slates to resist driving rain, or significant roof overhangs in areas with high rainfall, and try to replicate these functions in the detailing of the proposed building.

Vernacular buildings, such as this Derbyshire miner's cottage built about 1840, can provide lessons in how to site and orientate a new building, and how to make the best use of local materials.

Indicative detailing

The aim of this chapter is not to show details of, for example, *the* plinth or *the* eaves ready to transpose directly into your design. There are so many potential variations and combinations of design and materials that this would be impossible. Instead, we highlight some of the common problems and recurring themes encountered when detailing some of the more technical parts of the hempcrete building, and, where appropriate, present some functioning solutions.

An important concept central to many detailing decisions when designing any building is airtightness. Hempcrete buildings are no exception, and we therefore discuss the concept of airtightness first in this chapter, and return to it throughout as we look in more depth at detailing specific parts of the building.

Note: all the details described or shown in this book are for indicative purposes only. They should be considered as inspiration for your own specific building design, the actual detailing of which should be undertaken by a suitably experienced or qualified designer, architect or engineer. The overall design of any building, including individual details thereof, remains the responsibility of the building designer. A building's design should be considered for appropriateness within its own site and climate. If your architect or engineer is in any doubt about this, then seek specific advice from someone who is used to working with hempcrete.

Airtightness

Airtightness has become the buzzword of the moment in the field of sustainable building, as those people who want future-proof homes rush to seal themselves into plastic boxes in the name of reducing energy consumption.

Airtightness *is* a key concept, and a certain level of airtightness is needed in order to stop heat transferring out of the building via draughts in

Left: Hung tiles used as cladding over hempcrete give extra protection to exposed areas, such as on this gable end.

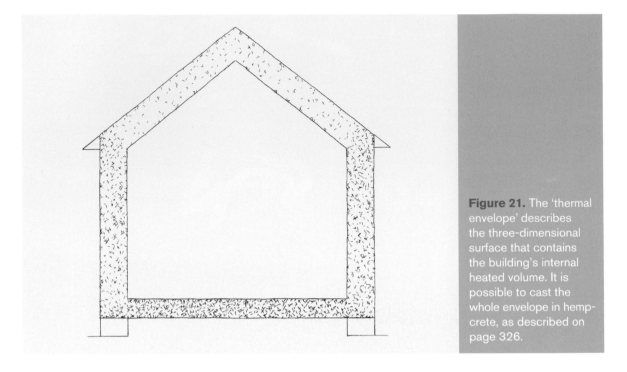

Figure 21. The 'thermal envelope' describes the three-dimensional surface that contains the building's internal heated volume. It is possible to cast the whole envelope in hempcrete, as described on page 326.

leaky buildings. The 'airtight line' is a conceptual line drawn around the thermal envelope (see Figure 21), along which all junctions of different building elements are detailed to ensure that air cannot escape between them. Where the building elements along this line are not in themselves airtight, they can be improved with an airtight membrane. In effect, the *conceptual* airtight line ensures that the building's designer has to consider the airtight envelope. Unfortunately, however, this widely used concept of the airtight line around the thermal envelope is often taken rather too literally.

This results in an arguably wasteful approach whereby the building is literally wrapped in an airtight membrane, taped at joins and around protrusions. In conventional construction methods this usually leads to the creation of a boarded-out void on the inside of the external walls, or even two – one to take the services and another to take the airtight membrane.

There are two problems inherent in this approach. The first is that it relies on an additional product in the wall build-up to achieve what could be achieved through thoughtful detailing. The second is that the placement of a physical airtight membrane, apart from being more expensive than simply using robust detailing, means that the airtight line is often on the interior face of the building. This means that the insulation, outside the airtight line, is exposed to cold air moving through it. A better approach is to position the insulation *inside* the airtight line, so that warm air is retained in the insulation material.

Hempcrete has an advantage over other building materials when it comes to airtightness, owing to its monolithic structure and its ability to be cast easily around protrusions and into different shapes. However, it is important to understand that it is the thickness and mass of the hempcrete, coupled with the density of the thin layer of render, that ensures good airtightness in a hemp-crete wall. Where cladding is used, meaning that an attractive render finish is unnecessary, the airtightness of the wall can be ensured with an airtight membrane or a rough basecoat render applied to the face of the hempcrete under the cladding.

All of this means that when considering the airtight line, both the hempcrete and the render (or membrane under cladding, if you are using one) must be sealed at all junctions with other building elements. To improve the airtightness between hempcrete and these other elements, consider increasing the area of the join and also introducing a change in direction to it. By increasing the area, you reduce the risk that a small failure of one part could cause a breach at the junction, i.e. making it less likely that an isolated failure would have any detrimental effect on the overall join. By introducing a change in direction, you are further obstructing the

A note on airtightness versus indoor air quality

The level of airtightness in hempcrete buildings should be considered within the particular context of how the material works compared with conventional timber-frame insulation materials. Conventional timber-frame insulation is usually very lightweight and low density, and the insulation is provided solely by the air trapped in the structure of the material. In the case of hempcrete, trapped air in the material provides the same function, but in addition hempcrete has a relatively high density, giving it good thermal mass. This allows a hempcrete wall to act the same way that a heat brick does in a wood-burning stove or storage heater: storing heat and radiating it out into the room long after the heat source (whether fire or electrically generated) has disappeared.

In a lightweight timber-frame house with low-density insulation, a high level of airtightness is essential because it is only the hot air trapped inside the building, by the insulation, that is holding the heat. Any air leakage swiftly causes a transfer of heat out of the building. In contrast, in a hempcrete building the thermal mass of the walls stores the heat *in the material itself* and slowly releases it again. Therefore, even if

warm air from inside a hempcrete building does escape, there is still further heat retained in the walls of the building.

This means that there is much more flexibility within a hempcrete building for natural ventilation, i.e. opening windows to allow exchange of air, rather than relying on mechanical heat-recovery ventilation systems such as those specified in high-tech super-insulated (but low thermal mass) buildings. In short, hempcrete is a combination of modern building elements (insulation through trapped air) with traditional ones (high thermal mass), and in many ways makes use of the best features of each technology to achieve its unique thermal performance.

Indeed, the potential of the heat store within the building created by the thermal mass of the hempcrete can be maximized by specifying hempcrete for the internal walls as well as the external ones. Using hempcrete internally also makes use of its excellent acoustic insulation properties to dampen the transfer of sound between rooms. For more information on the thermal and acoustic properties of hempcrete, see Chapter 7.

potential passage of air at the join. An example of such a solution is shown in Figure 29(c) (page 325), which illustrates airtightness detailing at the eaves.

As discussed in Chapter 21, the render can usually be sealed to other materials using a render stop bead and a suitable sealant such as burnt sand mastic or linseed oil mastic sealant, but the hempcrete will require an expandable material to take up any shrinkage (the material will also need to be able to withstand the wet environment if fitted against wet hempcrete). If a membrane is used in conjunction with cladding, it can be sealed with proprietary sealing tape or by compression when screwed between a batten and other building element (see page 331).

It is worth noting that the actual airtightness levels achieved by hempcrete in tests are very good, as illustrated by the following examples. During research carried out on the 'Hempod', an experimental hemp building with 200mm-thick walls built at Bath University:

> The air permeability of the Hempod was tested prior to the co-heating test in accordance with the procedures detailed in ATTMA TS1 [13] and BS EN 13829 [14] test Method A; thus the building was tested in its finished state with no temporary seals. The number of air changes per hour at 50Pa (n_{50}) was 0.55, which is within the Passivhaus limiting value of 0.6 air changes per hour[1]

An airtightness test on a timber-frame construction with hempcrete infill as part of the Serve Project in 2010, at Cloughjordan Ecovillage in Co. Tipperary, Ireland, showed "an excellent airtightness level of $1.12m^3(m^2.h)@50Pa$".[2]

Plinth

The main function of the plinth is effectively to be a 'good pair of boots' for the building, lifting the hempcrete walls at least 250mm above the outside ground level in order to protect them from splashing rain and groundwater.

The difficulty in plinth design lies in the conflicting functions required of it. The plinth needs to be strong enough to take any frame loading placed on it, yet still provide adequate insulation. It needs to be waterproof, yet not seal the hempcrete in among too many non-permeable surfaces (e.g. in the case of a breathable floor construction, as shown in Figure 26 (page 316), the hempcrete may continue down the inside face of the plinth, which thereby seals the hempcrete on its outside face). It is unlikely that you will find a material that is insulating, load-bearing, breathable and waterproof (though recycled glass foam comes pretty close). You therefore need to have a range of different materials doing different jobs in a relatively small place, which results in a technical challenge. Solutions to this problem have included employing multi-functioning or high-performance materials, or designing out one of the functions of the plinth.

Note that in Figures 22, 23, 25 and 26 on the following pages, which illustrate plinth detailing, floor heights are shown as in the most likely position; however, the exact floor height will vary dependent on a range of planning and design factors, for example the total maximum roof height allowed and/or the materials from which the plinth is built.

Multi-functioning materials

Recycled glass foam is becoming a popular solution to perimeter insulation throughout the construction industry. Its closed-cell structure prevents capillary action, i.e. it doesn't 'wick up' water as a lightweight concrete block would do. In addition, it has good compressive strength and can be used in structural applications, but is also low density and so provides insulation. Recycled glass foam can be used within the plinth construction, supporting the structural frame while simultaneously providing insulation. Like all materials, however, it does have limits in terms of its structural properties, and should therefore be specified by a competent person and in consultation with the manufacturer or supplier.

An example of recycled glass foam plinth detailing is shown in Figure 22, using engineering bricks at the external face and with a solid floor build-up: a free-draining design. The engineering bricks and NHL 5 mortar stop rainwater entering the

plinth, and the non-capillary-action sub-base floor layer and glass foam blocks prevent rising damp. Coupled with a free-draining foundation, this build-up would not require any PVC damp-proof course or membrane. A similar detail is pictured on page 153.

Plinth with no inherent insulation value

If the plinth is narrower than the total wall thickness, a material with high thermal perform-ance can be utilized to make up the difference in width. Although thinner than the hempcrete wall above it, a high-performing insulation should be capable of providing a similar level of insulation to the hempcrete. Unfortunately, many materials that suggest themselves for use

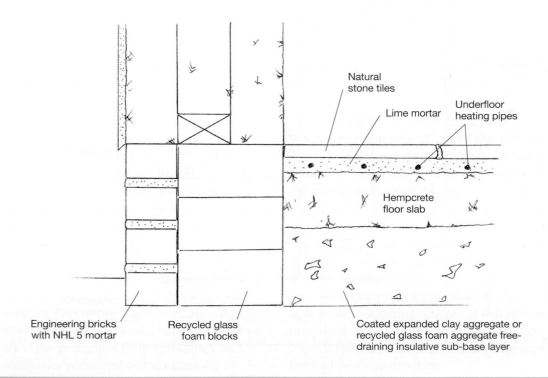

Engineering bricks with NHL 5 mortar

Recycled glass foam blocks

Natural stone tiles

Lime mortar

Underfloor heating pipes

Hempcrete floor slab

Coated expanded clay aggregate or recycled glass foam aggregate free-draining insulative sub-base layer

Figure 22. Recycled glass foam is probably the best load-bearing plinth insulation available currently.

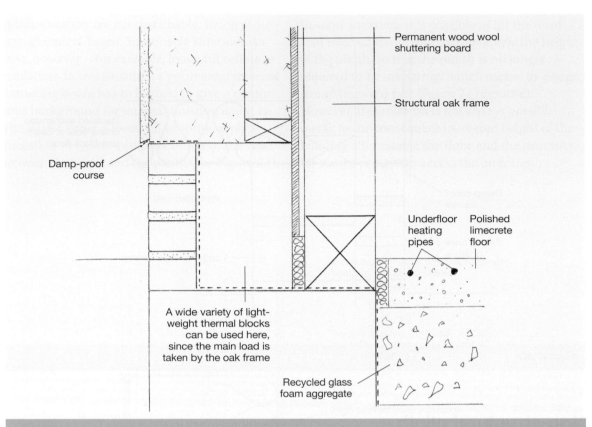

Permanent wood wool
shuttering board

Structural oak frame

Damp-proof
course

Underfloor   Polished
heating   limecrete
pipes   floor

A wide variety of light-
weight thermal blocks
can be used here,
since the main load is
taken by the oak frame

Recycled glass
foam aggregate

Figure 25. A separate structural frame (in this case oak) allows for a non-load-bearing plinth.

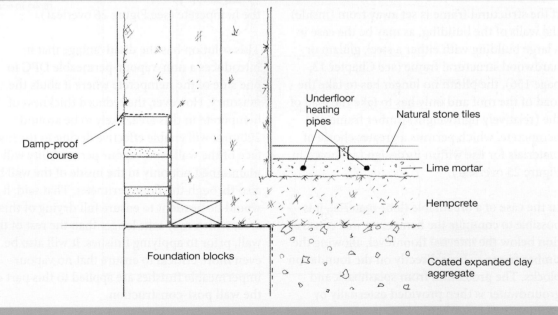

Underfloor
heating
pipes   Natural stone tiles

Damp-proof
course   Lime mortar

                                         Hempcrete

Foundation blocks

                                         Coated expanded clay
                                         aggregate

Figure 26. A breathable floor allows the timber frame to bear directly on the foundation blocks.

General design considerations

The plinth arrangement may affect, or be affected by, the position of the structural frame within the wall and, to a lesser extent, the position of the doors. The plinth design might only allow positioning of the frame to the inside, middle or outside of the wall, depending on the load-bearing capabilities of that section of the plinth construction, and so the structural frame and the plinth should always be considered together at the design stage.

The way in which the plinth interacts with door openings should also be considered carefully. In situations where the internal floor level is below the top of the plinth, the door frame will be meeting the plinth as well as the hempcrete wall above it. For this reason it is very important, at the design stage, to draw up a horizontal section of the door opening where it interacts with the plinth, as well as where it interacts with the hempcrete part of the wall. This may affect where and how the door frames are fixed, and thus the plinth detail at that location.

Airtightness at the plinth isn't as much of a problem issue as in other areas, since gravity acting on the mass of the hempcrete is compressing the junction between plinth and wall. However, there is a junction that goes from outside to inside across the DPC, and there remains a possibility of air moving along this line. If this is an issue, for example if a very slow-setting binder is to be used and there are concerns about the hempcrete shrinking away from the DPC, a seal can be formed between the DPC and the underside of the render stop bead with a linseed oil or burnt sand mastic. It is also possible to use a compressed hemp-fibre quilt insulation airtightness detail (similar to those described in Chapter 21, page 297) around the DPC. However, in reality this is very unusual, and in our own practice we

Plinths can be more sustainable than a concrete upstand with cement render, as shown by this plinth made from reclaimed brick and lime render.

have never seen a need to take specific airtightness measures at the plinth.

Hempcrete floors

Although widely used throughout France, hempcrete floors are less common in the UK, where the concept of a breathable – moisture-permeable – floor is largely an alien one. As discussed in earlier chapters, over the last 50 or so years there has been widespread use of inappropriate materials and detailing in the repair and restoration of properties dating from the Victorian period and before. This has caused countless instances of damp problems in floors and the lower parts of walls, since the traditional breathable solid-wall construction has been stopped from working properly. The cause of these damp problems has not been widely understood in the construction industry, and more often than not has been falsely linked with the original construction methods, including the use of lime mortars and traditional building techniques.

The UK building industry has therefore become suspicious of traditional methods of construction, and the prevailing attitude is that concrete and plastics are the only materials suitable to be in contact with the ground. The weight of this conventional wisdom has led to a very slow acceptance in the UK of the use of lime and plant fibres in floor construction, especially where a damp-proof membrane (DPM) is not used. However, if detailed correctly, hempcrete or limecrete floors can outcompete their cement-based counterparts in terms of thermal performance, and are the best solution for the restoration or upgrade of historic buildings.

Having said this, not all sites are suitable for hempcrete floors; in particular, sites with very high water tables or in flood-prone areas would not be suitable, because if the water table were to rise higher than the sub-base level, water could fill this layer and reach the hempcrete. A hempcrete floor would survive a one-off flood and dry out, but would not be able to tolerate frequent wetting. In any case it is normal to construct your hempcrete floor slab at a height above the outside ground level. Note that breathable floors are also not suitable for areas where a radon barrier is required in the floor.

Detailing of hempcrete floors is quite straightforward. The existing hard core is levelled and compacted and, if necessary, a layer of pea shingle can be used to fill and smooth any large gaps (e.g. if a large-sized aggregate such as brick rubble has been used). A non-capillary, insulating subbase (such as *coated* expanded clay aggregate or recycled glass foam aggregate) is the next layer. This effectively becomes the DPM, since moisture cannot move up through it. On top of this a breathable geotextile membrane can be used to keep the sub-base separate from the hempcrete, which forms the next layer (see Figure 27).

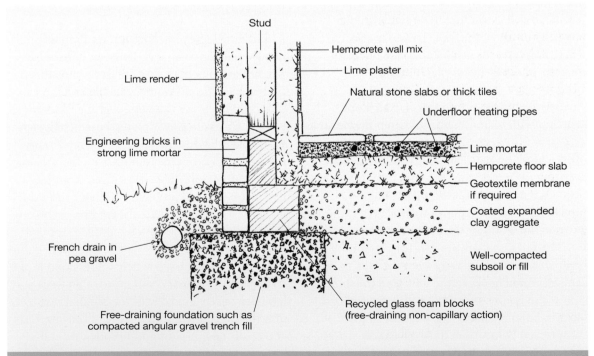

Figure labels:
- Stud
- Hempcrete wall mix
- Lime plaster
- Lime render
- Natural stone slabs or thick tiles
- Underfloor heating pipes
- Engineering bricks in strong lime mortar
- Lime mortar
- Hempcrete floor slab
- Geotextile membrane if required
- Coated expanded clay aggregate
- French drain in pea gravel
- Well-compacted subsoil or fill
- Free-draining foundation such as compacted angular gravel trench fill
- Recycled glass foam blocks (free-draining non-capillary action)

Figure 27. A 'breathable' floor build-up. The free-draining foundation and plinth, coupled with a vapour-permeable floor build-up, does not require a damp-proof course in the wall or a damp-proof membrane in the floor.

However, this is not strictly necessary if each layer is well consolidated and levelled and if care is taken during the application of each.

As discussed in Chapter 17, the hempcrete used in floors is a denser mix than wall hempcrete, so that it can withstand the typical loads expected of domestic floors. Possible finishes include solid timber (where a batten has been cast flush into the floor surface to provide a fix), natural flagstones set in lime mortar, or a lime screed with a natural stone or clay floor tile finish. Underfloor heating pipes can be included in the lime mortar layer for the flagstones, or in the lime screed layer for the natural floor tiles.

The thickness of these two main layers, hempcrete and insulating sub-base, depends on the desired U-value, which in turn depends on the dimensions of the floor, the quality of the perimeter insulation and the ground type. Generally, the hempcrete layer is worked out to be as thin as possible for the structure to be stable (typically between 80mm and 150mm), and the sub-base is then as deep as is needed to make up the U-value (typically between 120mm and 180mm). It is possible to satisfy the requirements of Building Regulations for a small extension structure with a total build-up of hempcrete and sub-base of 200mm. The exact thickness of the layers should be specified by the supplier, who can input dimensions of the floor, type of perimeter insulation, ground type, U-value required, etc. into a piece of software that will generate the depth of each layer required.

Since the hempcrete provides both the structure and part of the insulation in one go, whereas concrete provides only structure and requires a substantial amount of extra insulation, the depth of a hempcrete floor build-up will often be less than that of a typical concrete floor build-up, offering savings on excavation and waste costs.

Being a material with good flexural strength, hempcrete will not require expansion joints within the floor or at the junction with the plinth. However, if the designer was concerned about airtightness, then an expandable material can be placed between the hempcrete and/or screed and the vertical DPC within the plinth. It would also be possible to mechanically fix or tape the membrane in the floor build-up to the DPC, although this would incur the extra cost of a very strong, breathable, airtight membrane designed for use in the floor instead of the relatively cheap standard geotextile membrane.

When detailing a hempcrete floor, the DPC within the plinth construction (which cannot connect to the DPM, as there isn't one) should continue down the inside of the plinth, and foundation blocks if necessary, to at least the bottom of the sub-base layer. This will prevent lateral movement of water through the plinth into the hempcrete or sub-base layer.

All buildings with solid hempcrete floors should also include a French drain externally around the walls of the building, to assist in directing water away from the building. In addition, it is good practice to ensure that the ground in the immediate vicinity of the building slopes away from it whenever possible.

Structural frame (racking strength)

As described in the previous chapter, racking strength is the ability of a structure to withstand movement, for example from external stresses such as wind loading, without toppling over. Hempcrete provides a great deal of racking strength to the structural frame, but scientific data in this area is still quite limited, and many UK structural engineers and building control inspectors will remain unconvinced about the extent to which it can do this. Some methods by which extra racking strength can be provided are discussed as follows.

On large steel- or glulam-framed buildings, the frame usually sits inside the walls of the building, not within the hempcrete. The racking system will then have little interaction with the hempcrete and is likely to consist of diagonal frame members or steel ties.

On smaller builds of timber-frame construction, such as single dwellings or extensions, the frame is likely to be encased within the hempcrete, and as such the racking provision must not interfere with the vapour permeability and structural integrity of the hempcrete. For this reason, the usual method of adding racking strength to timber frames – OSB or plywood boards fixed to the frame within the wall build-up – is not

suitable. Such boards, if placed within the hempcrete, would cause a continuous vertical break midway in the mass of hempcrete and thereby weaken the monolithic structure. Even if placed on the inside or outside face of the wall, so as not to cause a break in the hempcrete, these boards are not vapour permeable enough to be left in situ. Plywood, being layers of timber glued together, is pretty much impermeable to moisture (and can easily be waterproofed with oil or resin). OSB is somewhat vapour permeable, but not to the extent that it could be used within a 'breathable wall' build-up.

This is not to say that OSB can't be used within a hempcrete wall for other applications. For example, it is used as the through-wall connecting element in a double-frame structure (see Chapter 13, page 160). The difference between this and using it as a racking board is that when used as a racking board on one side of the hempcrete,

Social housing at Callowlands in Watford, built using a timber frame and cast-in-situ hempcrete.

the OSB forms a complete barrier against the lateral flow of vapour, forcing any moisture to travel through it. When used as the joining plate within a double frame, it is positioned in line with the flow of vapour and so does not form a barrier to it. Moisture will take the easiest route from one side of the wall to the other, which will be through the far-more-permeable hempcrete rather than the OSB.

Viable methods for providing additional racking strength to the softwood frame within a hempcrete wall are: a vapour-permeable plaster carrier board, diagonal timber bracing, diagonal stainless-steel straps, or any combination of the three. These methods are discussed below. It is important to stress that the total racking strength required is dependent on many factors, including prevailing wind speeds, the loading of the building, the shape of the building, whether there are structural internal walls, etc. The strength required, and the means used to achieve it, should be calculated, given knowledge of the building design and the declared racking strength of the materials to be used, by a suitably competent person such as a structural engineer.

Vapour-permeable plaster carrier boards

The two options available at the time of writing are wood wool board and magnesium silicate board, as described in Chapter 14. Magnesium silicate board provides the higher racking strength of the two, and, having a very smooth surface, it requires only a skim of topcoat plaster to finish it. However, since this will usually be a gypsum skim it is not the ideal finish for a breathable hempcrete wall. The smooth surface of these boards also means they do not key to the hempcrete very well, and in our experience they seem to de-laminate from the hempcrete as it dries. This is corroborated by a 2013 evaluation of the Renewable House Programme, which notes that

"a further factor reported by builders was that the hemp and lime may have de-bonded from the Resistant [magnesium silicate] board permanent shuttering".[3]

Magnesium silicate boards are also made up of a very dense material which, while sold as vapour permeable, doesn't seem to have the same clear breathability characteristics in use as more open-structured boards do. We have noted large patches of damp and discolouration occurring on these boards, which remained for several weeks after the casting of the hempcrete. These were presumably caused by the moisture passing very slowly out through the board. Nor are these boards able to carry traditional lime plasters, due to their very smooth surface, and this can significantly reduce finishing costs on a hempcrete build. To our knowledge, at the time of writing magnesium silicate boards are all produced in India and China and so may have a high carbon footprint owing to being transported halfway around the globe. Very little information is available about exactly what goes into the boards, the quality standards of the products, or working conditions in the factories.

Wood wool boards have a very open surface structure, similar to that of hempcrete. Because of this the hempcrete keys very well to the board, leaving no voids. However, the open, rough surface of the wood wool board means that at least two skims or a basecoat and a topcoat will be required to finish the board. An alkali-resistant plasterer's mesh is usually incorporated into the basecoat to stop cracking at joins between boards. Some wood wool boards have the advantage of being external grade and so can be used internally or externally to carry the render. At the end of the day, the deciding factor in choice of carrier boards may be budget, with magnesium silicate board at the time of writing costing approximately four times as much as the most expensive wood wool boards.

Diagonal timber bracing

This is the usual option for providing racking strength to the frame, and for those who value the simplicity of a monolithic hempcrete wall, with finishes on both sides applied directly to the surface of the hempcrete, it is probably the best. These bracing timbers usually consist of pieces of timber 50mm x 100mm or 50mm x 150mm in section, or 150mm-wide strips of 25mm thick ply, screwed to one side of the frame diagonally. The size, type, quantity and spacing of the timbers should be specified by a suitably competent person. Diagonal timber bracing is sometimes used in conjunction with timber gussets: triangular pieces of plywood fixed to the junctions of the studs with the sole or wall plates. They add extra racking strength and also provide an excellent fix between the studs and the wall or sole plate, as an alternative to angle brackets.

Diagonal stainless-steel strapping

Not usually relied upon as the sole means of applying diagonal bracing to a frame, this consists of stainless steel that comes on a roll and is screwed to the frame diagonally. It is available in various widths and thicknesses and can be found in most builders' merchants. This strapping only works in tension, so it is important to understand which way you are trying to stop the building leaning when using it for racking strength. If in doubt, fix it diagonally in both directions.

Roof

The main considerations when detailing any modern roof, apart from waterproofing, obviously, are the requirements for airtightness and insulation. The use of cast hempcrete as roof

Timber frame with diagonal plywood braces and timber gussets.

Diagonal stainless-steel strapping used as a brace.
Image: Peter King

insulation, as described in Chapter 17, has the advantage of introducing increased thermal mass into the roof structure.

Eaves

The main function provided by good detailing at the eaves is the first defence against precipitation by means of a good roof overhang. The other important consideration with this detail in hempcrete buildings is how to form a robust junction between the wall hempcrete and the roof insulation that is also straightforward to construct and airtight – which can be difficult to achieve. As discussed in Chapter 21, it is also necessary to consider the construction process when detailing the eaves, with regard to the advisability of placing the hempcrete from above.

Airtightness at the eaves is more difficult to achieve than at other junctions. One of the reasons for this is that at the eaves of most buildings the rafters project through the airtight line, because the vapour-permeable roof membrane which in most pitched roofs is used on top of the insulation has to come down and join with the airtight element in the wall. This usually requires the time-consuming job of taping a hole in the membrane, rather clumsily (and not very effectively), around each of the rafters. This problem can be solved by using prefabricated bolt-on rafter ends, as they are fixed after the airtight line has been completed, meaning that the roof membrane can continue seamlessly down to the top of the wall (see Figure 28 overleaf). Bolt-on rafter ends also offer the following advantages to a hempcrete build:

- They do not get covered with hempcrete, as they are fitted after the hempcrete is installed. This is useful if the rafter ends are to be left exposed at the eaves (visible from the ground outside).
- Because they are absent when the hempcrete is placed, the rafter ends don't obstruct this

process as they would do otherwise.
- Because they are not a continuation of the main rafters, as would be usual, this allows the main rafters to be an engineered I-section timber while the rafter ends can be a solid timber. On large roofs this results in a cost saving, while still allowing aesthetically attractive exposed solid-wood rafter ends. These are often desirable, for example on attractive oak-framed buildings.
- They make it a simple process to reduce the pitch of the roof at the eaves (a standard detail on roofs with a medium-to-steep pitch, which helps to slow rainwater down as it approaches the guttering).
- The rafter ends are very easy to replace, which can be useful since they are most vulnerable to the weather and are prone to rot.

The main disadvantages to using bolt-on rafter ends are that:
- They are more complicated to build and they are a relatively recent innovation, so not all construction workers will be familiar with them.
- There is a limit to the weight of the roof covering that can be used, since the bolt-on rafter ends are inherently weaker than a continuous rafter; for example, they would not be able to support the weight of a green roof.

Bolt-on rafter ends will not be suitable for every building design, but in our opinion they bring so many advantages that they should be incorporated wherever possible.

Another situation in which airtightness at the eaves can be difficult to achieve occurs when a lightweight quilt insulation is used in the roof (also shown in Figure 28). The answer to this is to use an insulating wood-fibre sarking board to compress the quilt insulation against the top of the hempcrete wall, as described on the following pages.

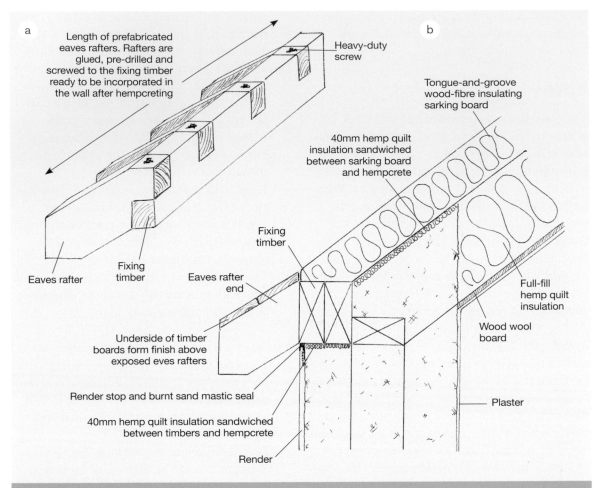

a

Length of prefabricated eaves rafters. Rafters are glued, pre-drilled and screwed to the fixing timber ready to be incorporated in the wall after hempcreting

Heavy-duty screw

b

Tongue-and-groove wood-fibre insulating sarking board

40mm hemp quilt insulation sandwiched between sarking board and hempcrete

Fixing timber

Eaves rafter

Fixing timber

Eaves rafter end

Full-fill hemp quilt insulation

Wood wool board

Underside of timber boards form finish above exposed eves rafters

Render stop and burnt sand mastic seal

Plaster

40mm hemp quilt insulation sandwiched between timbers and hempcrete

Render

Figure 28. Prefabricated bolt-on eaves rafter ends. (a) Rafter ends are attached to the fixing timber before fitting to the wall. (b) Rafter ends in the finished wall.

Airtightness over the roof

The vapour-permeable membrane generally used on top of the roof insulation provides a second line of defence against any moisture that penetrates the roof covering. Any moisture that moves out of the building via this membrane, which condenses as water droplets on the underside of the roof covering, is also then kept out by the membrane. The membrane, when joined with tapes designed specifically for the purpose, can also be used as the airtight layer in the roof build-up, and should be sealed to the hempcrete

and the render (or cladding membrane). The longevity of tapes in this situation is still largely untested, and in our experience it is very hard to get the overlap of membranes to lie perfectly flat for easy taping. For those who share our concerns, an alternative to using tape for airtightness at the roof membrane is shown in Figure 29 opposite.

The use of tongue-and-groove wood-fibre insulation boards for pitched roofs and walls is increasingly popular in the natural building world. Many of these boards perform the same function as a vapour-permeable membrane in terms of keeping water out while adding extra insulation. They are also more sustainable than such membranes, since although a processed material transported from Europe, they are produced from a natural material, wood, which acts as a carbon sink. Used as a sarking board, they can be installed over the rafters to provide a water barrier, and left without a final roof covering for up to three months while construction proceeds below. The tongue-and-grooved edge makes fitting easy, as the join between two boards doesn't have to sit above a rafter.

This join also gives a good level of airtightness in the roof, although some manufacturers also provide tapes to seal the joins. There are many different types of wood-fibre insulation boards available on the market, so it is important to make sure that weatherproof boards, specifically designed for this function, are used.

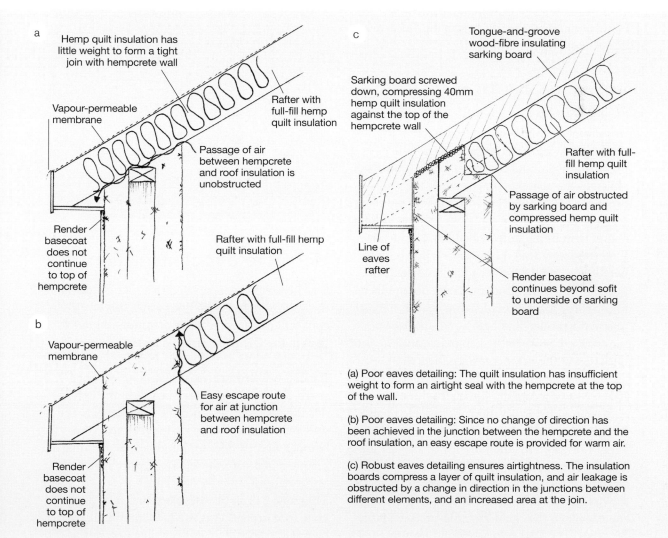

a

Hemp quilt insulation has little weight to form a tight join with hempcrete wall

Vapour-permeable membrane

Rafter with full-fill hemp quilt insulation

Passage of air between hempcrete and roof insulation is unobstructed

Render basecoat does not continue to top of hempcrete

Rafter with full-fill hemp quilt insulation

b

Vapour-permeable membrane

Easy escape route for air at junction between hempcrete and roof insulation

Render basecoat does not continue to top of hempcrete

c

Tongue-and-groove wood-fibre insulating sarking board

Sarking board screwed down, compressing 40mm hemp quilt insulation against the top of the hempcrete wall

Rafter with full-fill hemp quilt insulation

Passage of air obstructed by sarking board and compressed hemp quilt insulation

Line of eaves rafter

Render basecoat continues beyond sofit to underside of sarking board

(a) Poor eaves detailing: The quilt insulation has insufficient weight to form an airtight seal with the hempcrete at the top of the wall.

(b) Poor eaves detailing: Since no change of direction has been achieved in the junction between the hempcrete and the roof insulation, an easy escape route is provided for warm air.

(c) Robust eaves detailing ensures airtightness. The insulation boards compress a layer of quilt insulation, and air leakage is obstructed by a change in direction in the junctions between different elements, and an increased area at the join.

Figure 29. Good and bad airtightness detailing at the eaves.

The use of wood-fibre boards also assists in achieving airtightness at the eaves, for two reasons, as shown in Figure 29. First, it is very easy to introduce an obstruction to the passage of air at the airtight junction. Second, as described earlier, one of the main reasons that a good seal at the eaves is hard to achieve when using soft quilt roof insulation is that it is so lightweight that there is nothing to compress the seal between the insulation quilt and the top of the hempcrete walls. When fixing the boards down, a layer of quilt insulation can be trapped between the board and the hempcrete, squashing it down to give a much tighter seal.

Airtightness in a flat roof

Most modern flat roofs are constructed as a warm roof, with the insulation on top of the timber deck. Here the airtightness will be provided by the roof covering, which should be sealed to the airtight component in the walls, usually via the roof deck, with the roof covering glued to the top of the roof deck and the airtight component in the walls sealed to the underside. In a hempcrete wall this means sealing both the hempcrete and the render to the underside of the roof deck. The simplest way of doing this is to use some form of render stop bead and mastic to seal the render, and hemp quilt insulation sandwiched between the hempcrete and the roof deck to seal the hempcrete (see Figure 30).

In the case of a cold deck, the airtightness in the roof is provided by a vapour-permeable membrane on top of the hempcrete and under the vented air gap. This membrane can also form the airtight component in the wall, where a cladded finish is present, by continuing over the roof and down the wall, becoming the vapour-permeable membrane in the cladding build-up (see Figure 32, page 329). This cold roof build-up also allows the hempcrete wall to seamlessly become the hempcrete roof insulation with no join necessary.

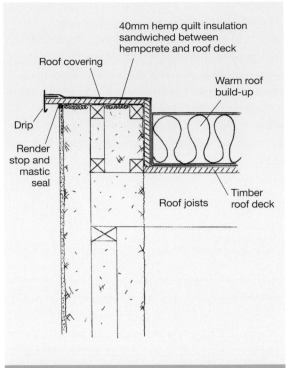

Figure 30. The airtight junction at the eaves of a flat roof with a warm roof build-up.

Labels: 40mm hemp quilt insulation sandwiched between hempcrete and roof deck · Roof covering · Warm roof build-up · Drip · Render stop and mastic seal · Roof joists · Timber roof deck

Hempcrete roof insulation

Cast hempcrete can be used as a roof insulation material, although, like floor slabs, this is an application that is far more common in France than in the UK at the time of writing.

If using cast hempcrete insulation in the roof space, the problem of airtightness in the eaves detailing is reduced, since the junction between wall and roof insulation no longer exists: the hempcrete continues up into the roof, forming a seamless join with the less dense roof hempcrete.

This concept can be extended further: if hempcrete roof insulation is used together with a hempcrete floor slab, it is possible to create a seamless 'box' of continuous cast natural insulation to provide an incredibly simple yet incredibly effective whole-thermal-envelope solution (see Figure 21, page 310). An extremely good level of airtightness can be achieved in this way, by designing out the presence of other materials and thus junctions and the use of sealants, tape and other 'weaker' elements. And, as discussed in earlier chapters, unlike synthetic whole-envelope systems, the hempcrete also provides a natural, breathable walling system with all the benefits this brings for the comfort and health of the building's occupants.

Although the use of hempcrete roof insulation can enable this elegant solution in terms of whole-building design, it has some disadvantages:

- It is heavier than most other roof insulations, and so may require a stronger roof construction than would otherwise have been required.
- As a mixed-on-site insulation, hempcrete involves an extra construction process compared with normal roof insulations, with associated additional labour costs.
- It has a lower U-value (at equivalent thickness) than fibre quilt insulation materials, and thus requires a thicker layer in the roof.

All three of these issues can be mitigated to some extent by the introduction of a layer of natural-fibre insulation (quilt- or board-type) above the hempcrete. This solution retains the advantages of the excellent airtightness and the thermal mass provided by the continuous cast hempcrete, while enabling a thinner layer to be used than would be required for hempcrete alone.

Hempcrete insulation in a pitched roof

Many options are available for insulating a pitched roof, depending on the type of rafter used, whether or not a hempcrete ceiling is desired (see page 330), and whether or not an insulating wood-fibre sarking board is to be used. Three such ways are illustrated in Figure 31

overleaf. In each illustration the tiling battens have been left out for the aid of clarity.

Down the length of the rafters, noggins will be needed to stop each newly filled section of hempcrete from sliding down the roof and compressing that at the bottom. (These are needed in all three options illustrated in Figure 31, but for simplicity they are shown in 31(b) only.)

The spacing of these noggins will depend on the pitch of the roof, with steeper roofs needing them at smaller intervals. These noggins can be thinner than the main rafters by 50mm at the top and bottom, in order to reduce any cold bridging effect. For example, if the main rafter were 200mm deep, the noggin could be 100mm deep, with its base 50mm above the base of the main rafter.

The simplest, and cheapest, method of insulating a roof with hempcrete is shown in Figure 31(a). To take full advantage of the loose-fill nature of the material, the main rafters have two sets of counter-battens, at least 50mm x 50mm section, fixed on top of them: the first horizontally and the second vertically in line with the rafters. This provides a fix for the horizontal tiling battens, and also allows the cast insulation to fully cover the rafters, minimizing cold bridging and also removing the potential air gaps right through the insulation along the side of the rafters. Between the counter-battens and the tiling battens a vapour-permeable membrane or wood-fibre sarking board is required as usual. If using a sarking board, however, it would only need to be 22mm thick for airtightness and waterproofing, which is much cheaper than 80mm boards.

The system of having layers of counter-battens sounds complicated, but actually the depth of a roof containing only hempcrete insulation needs to be at least 300mm in order for sufficient insulation to be achieved. Using a 300mm-deep rafter is expensive compared with using a 200mm

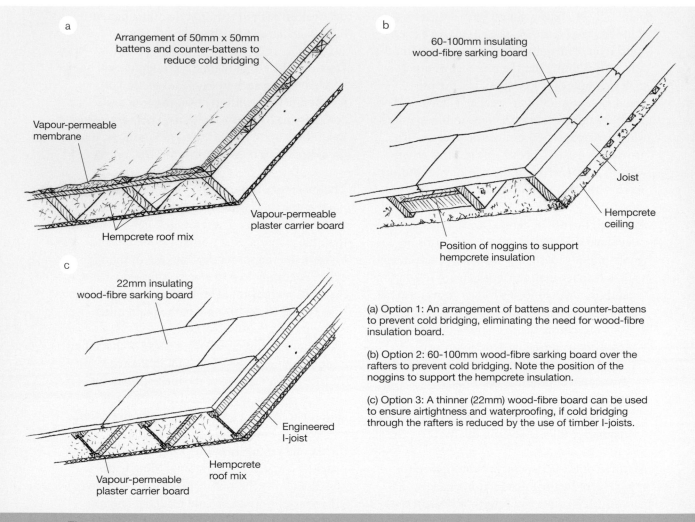

a

Arrangement of 50mm x 50mm battens and counter-battens to reduce cold bridging

Vapour-permeable membrane

Vapour-permeable plaster carrier board

Hempcrete roof mix

b

60-100mm insulating wood-fibre sarking board

Joist

Hempcrete ceiling

Position of noggins to support hempcrete insulation

c

22mm insulating wood-fibre sarking board

Engineered I-joist

Hempcrete roof mix

Vapour-permeable plaster carrier board

(a) Option 1: An arrangement of battens and counter-battens to prevent cold bridging, eliminating the need for wood-fibre insulation board.

(b) Option 2: 60-100mm wood-fibre sarking board over the rafters to prevent cold bridging. Note the position of the noggins to support the hempcrete insulation.

(c) Option 3: A thinner (22mm) wood-fibre board can be used to ensure airtightness and waterproofing, if cold bridging through the rafters is reduced by the use of timber I-joists.

Figure 31. Hempcrete insulation in a pitched roof.

rafter and two 50mm counter-battens, and the latter solution also results in a more robustly detailed roof in terms of maximizing airtightness and minimizing cold bridging. This system can also be used with other loose-fill insulations, for example recycled cellulose insulation.

The spacing of the counter-battens will depend on the weight of the roof covering they are supporting, and, along with the depth and type of the roof joist, should be specified by a suitably competent person.

Since cast-in-situ hempcrete in a roof is placed from above, there is an issue with keeping the work dry as it progresses, which applies to the walls too if these have already been cast. This can be achieved with a waterproof tarpaulin temporarily fixed at the top of the roof and rolled down each time work stops. However, fixing these coverings down and rolling them up when work starts again is a time-consuming and risky business, and on larger roofs the best solution is probably a scaffolding roof over the whole structure.

Keeping the work dry can be a challenge when casting roof insulation. *Image: Margot Voase*

Hempcrete insulation in a flat roof

When constructing a flat roof with cast-in-situ hempcrete insulation, a vapour-permeable membrane covers the top of the hempcrete, with a vented air gap above that (see Figure 32). To ensure airtightness and reduce cold bridging within the hempcrete layer, counter-battens can be used as they are in a pitched roof (see page 327). An air gap is achieved by running roofing battens on top of the vapour-permeable membrane and fixing the roof deck to these. Vents can be inserted into the fascia on either side of the roof to make the spaces between these battens into a series of air tunnels. Battens should be of good-quality treated or hardwood timber, owing to the potential exposure to moisture and the weight of the roof deck they are supporting. Once again, a suitably competent person should specify the spacing of the battens and the depth and type of the rafters.

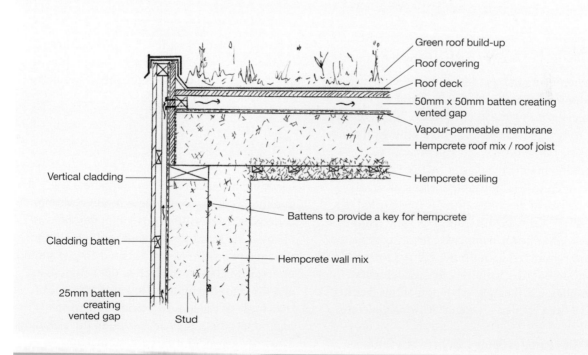

Figure 32. Hempcrete insulation in a flat roof.

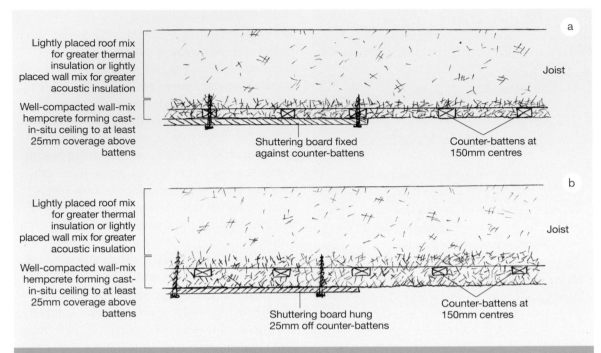

Lightly placed roof mix for greater thermal insulation or lightly placed wall mix for greater acoustic insulation

Well-compacted wall-mix hempcrete forming cast-in-situ ceiling to at least 25mm coverage above battens

Joist

Shuttering board fixed against counter-battens

Counter-battens at 150mm centres

Lightly placed roof mix for greater thermal insulation or lightly placed wall mix for greater acoustic insulation

Well-compacted wall-mix hempcrete forming cast-in-situ ceiling to at least 25mm coverage above battens

Joist

Shuttering board hung 25mm off counter-battens

Counter-battens at 150mm centres

Figure 33. A cast-in-situ hempcrete ceiling. Small-section timber battens are attached to the underside of the joists, and the hempcrete is cast around these. Shuttering board is either attached directly to the battens (a) or hung underneath them so the battens are not visible in the finished ceiling (b).

Hempcrete in ceilings

When using hempcrete roof insulation, it is important that the whole roof build-up is breathable, including the interior finish. If the roof space is to be used as a room, with a ceiling on the inside of the rafters, this can be achieved by using either a breathable plaster carrier board or a cast hempcrete ceiling coupled with a breathable finish. A hempcrete ceiling is constructed as a cast-in-situ structure from above and requires a layer of wall-mix hempcrete, firmly compacted and reinforced with timber battens to create a ceiling board. The lower-density hempcrete roof insulation is then placed on top of this (see Figure 33). See Chapter 17, page 226, for full details.

If the ceiling is being cast between floors or below a loft space rather than as part of a roof build-up, the preference may be for greater acoustic insulation rather than greater thermal insulation, in which case the mix used directly above the ceiling, between the joists, should be a (higher-density) wall mix.

Cladding

Cladding materials used on hempcrete walls, and the detail for fixing these to the cladding battens or frame timbers, is much the same as in any other building. As with most cladding systems for any building, a vented air gap should be created between the face of the hempcrete and the cladding matter. However, because of the nature of the hempcrete itself there are additional issues to be addressed in this detailing. The main considerations are as follows:

- The vapour-permeability of the hempcrete wall must be maintained.

■ If an externally exposed frame detail is used to aid fixing of the cladding, these frame timbers are vulnerable to moisture ingress and this issue needs to be resolved.

■ Since airtightness is not ensured in the absence of a render finish on the hempcrete, other measures need to be taken to maintain airtightness on the face of the wall.

The cladding by itself, with a vented air gap, provides sufficient protection for the hempcrete, but the requirement to protect the frame timbers and achieve extra airtightness presents a choice of solutions, with varying associated costs. This is an evolving science, with a variety of details in use and other possibilities that have not yet been proven. The full range of options is too extensive to discuss in detail here, but the three example options outlined on the following pages show different ways of resolving the issues of frame protection and airtightness.

Non-masonry cladding

As with any other timber-frame building, non-masonry cladding for a hempcrete wall, which may include timber, other boarded cladding or hung slates or tiles, is fixed to cladding battens that are situated within a ventilated air gap. The simplest way to ensure protection of the frame timbers and to achieve extra airtightness is to place the structural frame flush with the external face of the wall and fix the cladding battens directly to this, with a taped vapour-permeable membrane sandwiched between the two. The air gap between the cladding and the membrane, in which the battens sit, is vented at both ends to ensure a good flow of air behind the cavity (see Figure 34(a) overleaf).

This option has the advantages of both low cost and simplicity of construction, while achieving the key principles of ensuring airtightness and protecting the frame. However, the following

'cons' should be considered:

■ Although protected by the membrane, the external face of the wall is still a vulnerable position in which to place a softwood structural frame. Replacement of this section of the frame with hardwood would solve the problem, but increase costs.

■ The use of a vapour-permeable membrane introduces highly processed, high-embodied-energy synthetic materials into the building, which increases the environmental impact.

■ The effectiveness of using a taped vapour-permeable membrane to ensure airtightness in the long term depends very much on the quality and longevity of the membrane and tape, as well as on the thoroughness with which these are applied on-site.

A second, more sustainable, option avoids the need for a synthetic membrane by using a render coating to ensure airtightness under the cladding (see Figure 34(b)). Renders are not usually applied under cladding in conventional construction, since the material underneath is not usually capable of carrying a render, but hempcrete provides the ideal key for render. The costs are cheaper than for a standard external render, since a single rough basecoat is sufficient to provide airtightness and will not be seen under the cladding, so an attractive finish is not required.

While the render provides airtightness, it is not considered sufficient protection for a timber frame immediately below it, so the softwood structural frame is moved 50mm back into the wall. Normally a coverage of 70mm hempcrete would be recommended to protect timbers, but in this case, with the extra protection afforded by the cladding, 50mm is ample.

Placing the frame 50mm inside the hempcrete wall means that the cladding battens will not be fixed directly to it and therefore cannot transfer the load of the cladding to it. The battens are still

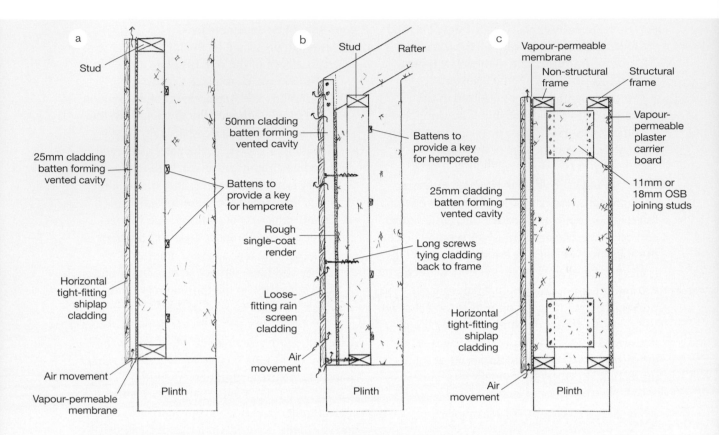

(a) Simple timber cladding detail. The structural frame is placed flush against the external face of the wall; airtightness and protection from moisture ingress are provided by a taped vapour-permeable membrane fixed across the frame.

(b) Natural cladding detail comprising a timber rain screen fixed to vertical cladding battens, which sit within the 50mm vented cavity and are fixed to the rafters and bear on the plinth. The structural frame is placed 50mm inside the wall and a rough basecoat of lime render is applied to the face of the hempcrete.

(c) Double-frame cladding detail, in which timber cladding is attached to a softwood cladding frame placed flush with the surface of the wall under a vapour-permeable membrane. The structural frame is on the internal face of the wall and therefore not vulnerable to moisture ingress.

Figure 34. Non-masonry cladding details.

fixed to the frame, with long screws through the hempcrete to hold them against the face of the wall, but in addition they need a structural fix to the rafters and also need to bear on to the plinth. Since the cladding battens are now taking the load of the cladding, they need to be larger in section than they would be if fixed directly to the frame. The size of these timbers should be specified by a suitably competent person, who should also consider the load of the cladding on the rafters (via the cladding battens).

The extra protection provided by the render may also allow the use of a simple rain screen cladding rather than a conventional tightly fitting or overlapping cladding. A rain screen cladding is constructed from boards that only need to butt up to each other, allowing the use of reclaimed timber instead of the usual featheredge or shiplap board.

A third option is to use a vapour-permeable membrane, as in Figure 34(a), but to improve the frame detail by using a double frame (see

Figure 34(c)). This involves moving the soft-wood structural frame to the internal face of the wall, where it is well protected, and connecting a separate non-structural cladding frame to it (see Chapter 13, page 160), which sits on the external face of the wall immediately under the membrane.

Since in this case the cladding frame only supports the weight of the cladding itself, rather than the load of the upper floors and roof (as in Figure 34(b)), it may be possible to construct it from smaller-section timbers (as specified by a suitably qualified person) than the structural frame. This in turn may make the use of hardwood a more viable option for this section of the frame, which further improves the detail. Weighed against this, however, are the additional material and labour costs of a using double frame, and the fact that a

synthetic vapour-permeable membrane should still be used to provide airtightness, although in the case of a hardwood frame it would not be needed to protect the timber. One of the biggest arguments in favour of the double-frame solution is the reduced labour and material costs on the shuttering, since temporary shuttering boards can be screwed straight to the frame on both sides.

From the examples set out here, it should be clear that the detailing of various claddings for hemp-crete is an evolving 'conversation' rather than a hard-and-fast set of rules. As long as the basic principles of structural integrity, airtightness and protection of surface timbers are addressed, there is no 'wrong' way of detailing it. The details discussed here are not intended to provide the reader with an exhaustive list of options; rather

Timber cladding over a hempcrete section in a cob wall.

Hung-tile cladding provides extra protection to this exposed hempcrete gable. This replicates the original gable detail in this listed timber-frame cottage.

to show some ways in which the issues can be addressed, while demonstrating that the way in which you deal with one issue can have an impact on the others – making the process quite complex and demanding of a holistic solution.

Masonry cladding

When applying masonry cladding to a hempcrete wall, the cladding bears on to the plinth and is tied back to the timber frame with stainless-steel cladding ties. As with other forms of cladding, a ventilated air gap is included (see Figure 35). Owing to the weight of masonry cladding and therefore its requirement for good support, it is sensible to keep the cladding as close to the timber frame as possible, which means the frame should be on the external side of the wall, with a vapour-permeable membrane fixed to it for airtightness and protection against moisture ingress. In this situation it is vital that the cavity between the membrane-covered frame and the masonry provides *effective* ventilation, so it is essential

that it is sufficiently wide and well vented, and does not get filled with dropped mortar. As many vents as possible should be provided at the base and top of the cladding.

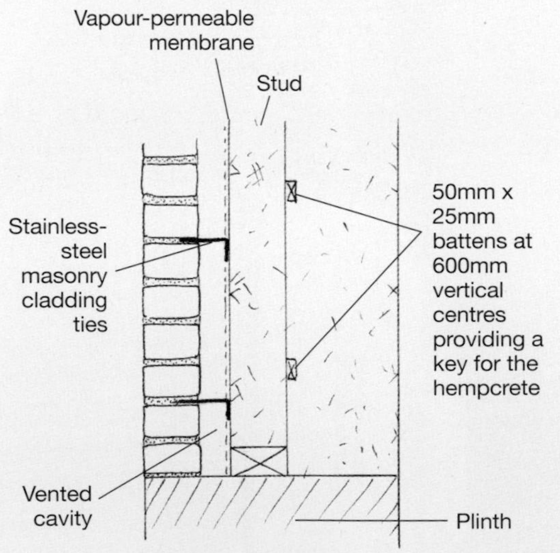

It is recommended that a lime-based mortar be used for masonry cladding, even *with* a vented air gap, as this has several advantages:

- This is a more robust detail in terms of breathability: a lime mortar helps to keep the masonry cladding permeable to vapour, which improves the overall breathability of the building, and allows any moisture in the cavity to exit through the cladding as well as through the vents.
- Lime mortar is more flexible and better equipped to absorb the slight movements that occur in timber-frame buildings.
- It is a more environmentally sustainable option.
- It will be more in keeping with the overall design and finish of the building.
- It will allow the use of softer stones and brick without causing damage to them.

Different types of building lime can result in wide variations in a mortar's hardness, tensile strength and vapour permeability, as well as in its other characteristics. When specifying a lime mortar for masonry, the main principle is that the mortar should be softer, in its set state, than the stone or brick used. This is so that any moisture taken in through the surface of the wall can easily be released through the 'sacrificial' mortar joints, rather than being held in the masonry where it has the potential to cause damage. Lime mortars should always be specified by a suitably qualified or experienced person who is used to using lime in masonry work.

The main issue when specifying a mortar for masonry cladding is that the mortar is both softer than the surrounding masonry while also strong enough for what is likely to be a relatively

Figure 35. Protecting frame timbers and ensuring airtightness if using masonry cladding.

Labels in figure:
- Vapour-permeable membrane
- Stud
- Stainless-steel masonry cladding ties
- 50mm x 25mm battens at 600mm vertical centres providing a key for the hempcrete
- Vented cavity
- Plinth

Masonry cladding on a hempcrete building. Note the external frame enabling the cladding to be tied in to the wall, and the vapour-permeable membrane.

thin masonry veneer. For such a thin depth of masonry cladding, bricks or a well-dressed or cut stone are likely to be the material of choice, for the increased stability their even shape brings.

Masonry cladding with no vented cavity

One possibility that many people are discussing currently, but which has not yet been proven in practice, is the idea of casting hempcrete directly up against masonry cladding, with no cavity. This could be advantageous because, where a cladding is to be used anyway for aesthetic reasons, this method simplifies the process and allows cost savings: on labour, as the external temporary shuttering is not needed, and on materials (e.g. the membrane and vents for the cavity).

To explore the effect this would have, it helps to compare the masonry cladding to a vapour-permeable lime render, which is the standard finish for a hempcrete wall. Imagine the masonry cladding as a very thick lime render (the mortar) with huge pieces of aggregate set in it (the bricks or stones). The problem is that this 'giant-sized aggregate' effectively creates breaks all the way through the render from the outside to the inside along the edges of the brick or stone.

There is a risk that water could move by wicking (capillary action) along the join between the brick and the mortar, straight through the 'render' and into the hempcrete. The hempcrete, being breathable, normally releases any moisture to the outside through the render once the rain has stopped. However, in this case, due to the wicking, there could be more moisture than normally expected, and an additional risk that some of this will get trapped behind the stones or bricks, which should be less vapour permeable than the mortar.

For this reason, it may take longer for the moisture to be released to the outside through the cladding than through a standard lime render. If the moisture hasn't been released by the time it rains again, then there could be a gradual build-up of moisture in the wall over time, to levels that the hempcrete would not be able to tolerate.

To counteract this effect, it may be possible to use a hydrophobic (water-repellent) yet vapour-permeable render on the inside face of the masonry prior to casting the hempcrete. Such products are available from manufacturers of high-tech proprietary lime-based renders, for example Baumit or Parex. This would stop any moisture droplets that wicked through the cladding from entering the hempcrete, but would still allow moisture vapour to pass in and out. Unfortunately, the cost of such renders is currently very high compared with that of a standard lime render, and as such might outweigh any savings gained by the elimination of the air gap.

The solution of using a hydrophobic render also presents problems for hempcrete drying, since there is liquid water as well as vapour which needs to dry out of freshly cast hempcrete. With a hydrophobic render already applied, then only the vapour could exit through the external cladding. All liquid water would have to dry to the inside of the hempcrete, which would happen eventually, but this would slow the drying time significantly. It is worth remembering, of course, that the simple presence of a masonry cladding directly against one side of the hempcrete wall will already have a huge impact on the drying time of the hempcrete and could potentially cause delays in the application of internal finishes, depending on the thickness of hempcrete used.

In addition, if moisture *vapour* can enter through the cladding, there is a risk that any vapour that subsequently condenses inside the hempcrete would not be able to leave again through the hydrophobic render. Again, this carries the risk of a build-up of moisture in the hempcrete.

This idea, then, requires more investigation, and trials in practice. It may be that where the wall is in a very sheltered position, the detail of hempcrete cast against masonry cladding *without* a hydrophobic render could work. This would be based on an assumption of limited weather exposure, and plenty of time between heavy soakings for the wall to expel all the water from the last rain. But if the wall was subject to even average exposure to rain, except perhaps in very dry climates, such a detail would certainly be inadvisable. In exposed locations, frequently subject to rain and high winds, this detail would not work, and indeed a more appropriate material in such a case might be a hung slate cladding.

Openings

The main considerations at openings in the hempcrete wall, apart from the familiar question of airtightness, are the need to provide a secure fixing for the window or door frame and to ensure that the hempcrete is well supported above the opening.

Fixing windows and doors

Window and door frames need to be fixed to the structural frame or to something that is strong enough to take the window or door frame fixings back to the structural frame.

Where the structural frame sits in line with the desired position of the window or door, the fixing can be achieved with little difficulty. In situations where the two are not in line, the detailing must provide a secure fix for the window or door frame. This can be done by boxing out with plywood or studwork and covering this with wood wool board to reduce cold bridging and provide a substrate for renders and/or plasters if required. However, where the windows sit towards the outside of the wall and the structural frame is positioned on the internal face of the wall, a straight boxing-out would create a deep 'tunnel' window reveal that cuts out light due to the depth of the wall. A more tapered or even a curved reveal can disguise the depth of the wall and increase the amount of light that can enter. This is achieved by stepping the frame and tapering the shuttering: the reveals can be left tapered, or brought to a curve with a nail float once the hempcrete has been placed (see Figure 36 opposite). It is more difficult to taper the reveals if a permanent internal shuttering board is being used, but if a 15mm wood wool board is used for the reveals this can be bent to form the taper and fixed to the studs as required.

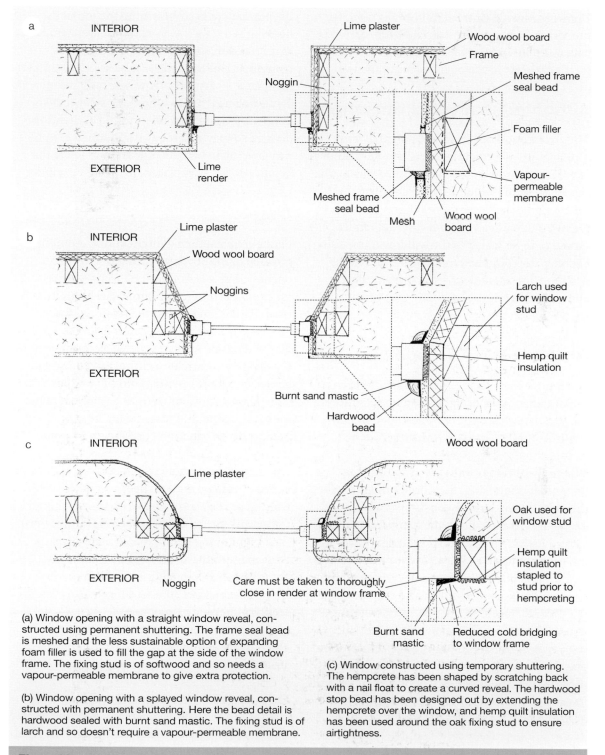

a   INTERIOR

Lime plaster

Wood wool board

Frame

Noggin

Meshed frame seal bead

Foam filler

Vapour-permeable membrane

Meshed frame seal bead

Mesh

Wood wool board

Lime render

EXTERIOR

b   INTERIOR

Lime plaster

Wood wool board

Noggins

Larch used for window stud

Hemp quilt insulation

EXTERIOR

Burnt sand mastic

Hardwood bead

Wood wool board

c   INTERIOR

Lime plaster

Oak used for window stud

Hemp quilt insulation stapled to stud prior to hempcreting

EXTERIOR   Noggin

Care must be taken to thoroughly close in render at window frame

Burnt sand mastic

Reduced cold bridging to window frame

(a) Window opening with a straight window reveal, constructed using permanent shuttering. The frame seal bead is meshed and the less sustainable option of expanding foam filler is used to fill the gap at the side of the window frame. The fixing stud is of softwood and so needs a vapour-permeable membrane to give extra protection.

(b) Window opening with a splayed window reveal, constructed with permanent shuttering. Here the bead detail is hardwood sealed with burnt sand mastic. The fixing stud is of larch and so doesn't require a vapour-permeable membrane.

(c) Window constructed using temporary shuttering. The hempcrete has been shaped by scratching back with a nail float to create a curved reveal. The hardwood stop bead has been designed out by extending the hempcrete over the window, and hemp quilt insulation has been used around the oak fixing stud to ensure airtightness.

Figure 30. Horizontal sections of window openings.

Unless hardwood or treated timber is used, the timbers around openings should be protected with a vapour-permeable membrane, owing to their proximity to the outside face of the wall.

Airtightness around openings

The area around openings is a potential weak spot in terms of airtightness. The window manufacturer's recommended methods for closing the gap between the window frame and structural timber frame should be followed, and, in addition, the internal plasterwork and external render should both be sealed to the window frame using a render stop bead and mastic seal, or a frame seal bead, the latter of which is aesthetically far more discreet for use internally (see Chapter 18, page 245).

Shuttering for an opening

Openings can be created using either temporary or permanent shuttering. If permanent shuttering boards are used for the walls, then it makes sense for these to be used for the reveals also.

Where the internal walls are created using temporary shuttering, then doing the same for the reveals gives a more consistent background for the plaster and allows the hempcrete to be curved with a nail float after placing. However, the use of temporary shuttering for the reveals requires the placing of the hempcrete around the openings to be very thorough and even, so that the corners created in the cast hempcrete are strong and stable. This takes extra time, since it is an awkward space to physically get into when placing the hempcrete. For this reason, those inexperienced with hempcrete building may wish to use permanent shuttering to make the placing of the hempcrete at this section less critical.

Permanent shuttering also makes it easy to support the hempcrete above the opening, since the permanent board forming the top of the opening is in place to hold up the wet hempcrete as it dries out through the board (whereas temporary shuttering has to be removed for the hempcrete to dry out). However, most permanent shuttering boards are inherently weaker than temporary shuttering boards, due to their lower density, so may still require additional temporary bracing to be fixed underneath while the hempcrete is being placed, to avoid movement.

Lintels

Permanent shuttering might at first seem the obvious solution when casting the top of an opening, but it is not difficult to provide support for drying hempcrete if temporary shuttering is used. Battens laid across the opening, spaced 50mm apart and at least 50mm from the external face of the wall, are cast into the hempcrete above the opening to provide reinforcement, extending across the whole opening into the wall on either side. These are either fixed to the frame or laid loose inside the cast material, bearing on the hempcrete on each side (see Figure 37 opposite).

Openings that do not have frame loads (joists or rafters) bearing on them need only the framework to provide a fix for the door or window frame and extra timbers as necessary to support the weight of the hempcrete. For short spans, these extra timbers might just be battens as described above, in addition to the door or window framework, but for longer spans, larger-section timbers or a separate structural lintel may be required to support the hempcrete. Openings bearing frame loads also require a structural lintel within the frame.

For extra structural integrity in the cast hempcrete at this vulnerable area, the lift of hempcrete that goes over the opening (and extends beyond each side of the opening) is placed in one go to form a cast-in-situ 'lintel' of hempcrete.

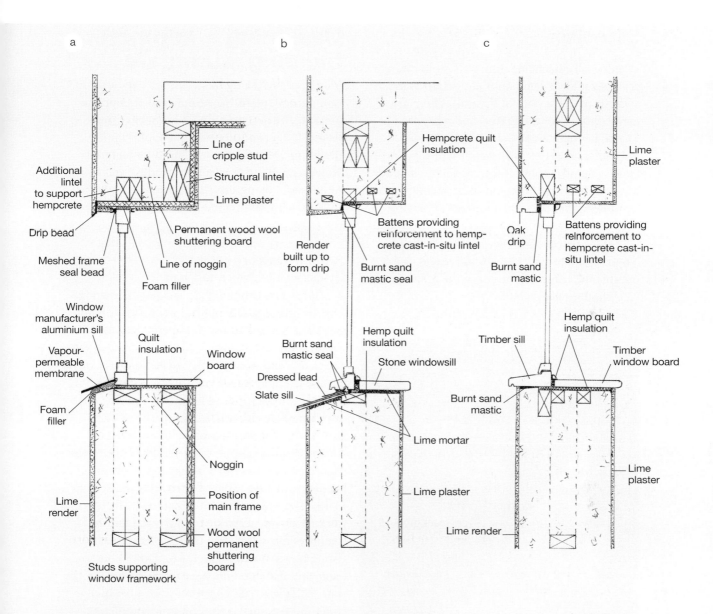

a

Additional
lintel
to support
hempcrete

Drip bead

Meshed frame
seal bead

Window
manufacturer's
aluminium sill

Vapour-
permeable
membrane

Foam
filler

Lime
render

Line of
cripple stud

Structural lintel

Lime plaster

Permanent wood wool
shuttering board

Line of noggin

Foam filler

Quilt
insulation

Window
board

Noggin

Position of
main frame

Wood wool
permanent
shuttering
board

Studs supporting
window framework

b

Hempcrete quilt
insulation

Battens providing
reinforcement to hemp-
crete cast-in-situ lintel

Burnt sand
mastic seal

Render
built up to
form drip

Burnt sand
mastic seal

Dressed lead

Slate sill

Hemp quilt
insulation

Stone windowsill

Lime mortar

Lime plaster

c

Lime
plaster

Battens providing
reinforcement to
hempcrete cast-in-
situ lintel

Oak
drip

Burnt sand
mastic

Timber sill

Burnt sand
mastic

Hemp quilt
insulation

Timber
window board

Lime
plaster

Lime render

(a) A wide-span window opening
on a building with an internal frame,
showing permanent shuttering and
additional lintel for the hempcrete,
together with supporting studs for
the window itself. Note the aluminium
windowsill and drip detail created
using a drip bead.

(b) A normal-sized window opening
on a building with a central frame,
constructed using temporary shuttering. Reinforcing battens are cast into
the hempcrete above the window to
support it. Note the natural slate and
lead sill detail, and the drip detail
created using render.

(c) A window opening with the same
construction as (b) but showing a
timber sill and drip detail.

Figure 37. Vertical sections of window openings.

Sill details

Windowsills in a hempcrete wall should have a good-sized drip detail on them to take water away from the surface of the wall. There are many options for doing this, and airtightness should be maintained using the methods described on page 338 and shown in Figures 36 and 37. Standard aluminium or timber sills can be used (but, depending on the placement of the window within the wall, a longer sill may be needed (see Figure 37(a) and (c)) or a secondary sill can be placed below it (see Figure 37(b)). Natural slate provides an attractive and durable option, and can easily be cut to the required size (see photo below).

Natural slate makes an attractive and effective sill detail.

Designing a hempcrete building: summary

As a natural, plant-based material, hempcrete presents the designer with a number of new factors to consider that do not apply when designing for conventional building materials. However, with proper understanding and a little forethought, hempcrete is a rewarding material to work with; its adaptability means that scope exists for a wide variation in different designs, and there are many different materials that perform well alongside it.

It is hoped that the indicative details shown in this section, together with the information and advice in the rest of this book, will provide inspiration to building designers and assist them in understanding not only the range of options available but also the consequence of these choices for the finished building, in terms of performance, cost and longevity.

In detailing a hempcrete building there are four main themes that the building designer must understand, as discussed in the previous chapter. These are: the limits of the material, in terms of its performance and structural integrity; the nature of the material, in terms of its key characteristics; how it behaves during and after construction; and how it will behave in the context of the particular site. The chance to visit a hempcrete building site during the construction phase, including seeing some unfinished but set hempcrete, will give the designer a real sense of the material and will provide a much more thorough understanding of the issues than a book could ever hope to convey. While this is not always possible currently, it should be much easier as hempcrete builds become more commonplace.

Some essential questions to ask when detailing and designing for hempcrete are:

- Does the design make sense within the structural limits of the material?
- Can it be constructed easily?
- Is the softwood structural frame safe from water ingress?
- Can all the materials and fixings used withstand the alkaline environment?
- Is vapour permeability maintained?
- Is airtightness maintained?
- Does the design reflect the demands of that particular building, on that specific site?

We recommend that, wherever possible, designers avoid complicating additions to the basic wall design, as the real beauty of hempcrete is in its simple, low-tech approach, with a minimum of specialized technical solutions required. Even in the use of permanent shuttering boards, for example, something intangible is lost. There is true elegance in the simplicity of the basic monolithic hempcrete wall with a central timber frame, and a lime render and plaster finish.

Once the key issues are understood, the process of designing a hempcrete building should be a relatively straightforward and rewarding one. The materials and methodology are low-tech and the process of building is simple. Architects who are involved in designing sustainable buildings and want these to be realized in a truly low-impact way, as well as those who are influenced by the traditional vernacular and seek the beauty of a simple, natural finish, will find much of value in hempcrete.

The structural frame for a new-build hempcrete extension. Note the glulam ridge beam and temporary bracing put in place to support the frame while it is being built.

Focus on self-build 3: Bridge End Cottage

When Bill met Jo Smith, she was living in a two-up-two-down brick 1880s Victorian cottage. When the house next door (the mirror image of Jo's, but with two large extensions added to the rear and the side) came up for sale, Bill bought it. The couple, together with Bill's young daughter Mollie and Mollie's grandfather Dave, then set about knocking the two buildings together.

The first thing that strikes you on walking into Bridge End Cottage, set in a quiet valley in the rolling North Wales countryside, is the sheer size of the home that Bill and Jo have created. Tardis-like, one room gives way to another . . . and yet another. Most of the individual rooms are large, and the light colours and large windows at the rear of the property give a feeling of light and space. The place is huge, and the mind boggles at how long the project must have taken them.

"It's been five years up to this point," says Jo, "and we've still got a little way to go." Almost every-thing about the house was replaced or renovated along the way: new roofs, a complete refurbish-ment of both properties and alterations to the layout of the internal spaces – the rooms on both floors, together with the connecting corridors and staircases. This got rid of any impression of 'two houses knocked into one' and made it feel like a single house with a layout that makes sense.

The poorly constructed kitchen extension on the rear of the second house was demolished and rebuilt with a green oak frame and hempcrete walls set outside of this. Hempcrete has also been used to insulate a flat roof, and to add breathable solid-wall insulation externally to the brick walls of the draughty Victorian properties, achieving not only a high standard of insulation but also tying together the 'mish-mash' of Victorian brick, twentieth-century add-ons and Bill and Jo's new extension, giving a consistent appearance to the whole house externally.

Bill and Jo were "sold on hempcrete right from the beginning" because they wanted to create a well-insulated home. Bill's brother-in-law Peter is an architect interested in sustainability, and helped to get Bill "fired up about insulation". Bill liked the idea of hempcrete, and had read some studies reporting that its performance in situ was significantly better than expected, due to its thermal 'buffering', whereby temperature changes are transferred through the walls so slowly that a constant comfortable temperature is maintained.

The other attraction of hempcrete was that "it was a low-tech job we could do ourselves, and didn't require any complicated equipment". Peter had rightly suggested that external insulation would be the best way to insulate the draughty Victorian property, since it would keep the thermal mass of the existing brick walls inside the 'tea cosy' of the hempcrete insulation. Bill developed a system of home-made ply I-joists nailed against the house wall which could have temporary shuttering fixed to the outside. 125mm hempcrete was then cast between the brick wall and the shuttering.

A lightweight Baumit lime-based coloured render containing polystyrene balls for extra

1. Bridge End Cottage, still warm in the snow. *Image: Bill Smith*
2. The kitchen in the new hempcrete extension.
3. The house is full of clever details, such as this window drip created within the render itself.
4. Upstairs in the hempcrete extension: the master bedroom.
5. This wood burner, together with heated towel rails upstairs, is the only heating the Smith family have used to date.

insulation was applied externally, and lime–sand plasters and breathable paints internally. A glass-fibre mesh was integrated into the render across the external surface to stop it cracking over the wooden I-joists visible on the surface.

For the rebuild of the kitchen extension, a green oak frame was erected and an additional 'four by two' softwood studwork frame offset from this to support the hempcrete. Internally, a wood wool permanent shuttering board was used. The next stage of the ongoing project will be the detached garage, which will also have hempcrete walls.

Bill and Jo's driving principle in the build has been thermal performance, and Bill's engineering background shows in the technological solutions they have found. They have not stuck exclusively to natural materials – for example, using renders containing polystyrene, and foam insulation boards for the insulation at rafter level in the warm roof. However, the oak structural frame in the new extension and oak second-fix carpentry throughout, together with the lime plasters, gives the house a very natural feel. Bill and Jo were keen to ensure a high level of airtightness (they commissioned and passed an airtightness test before they had finished the building!), and therefore ventilate the house with a mechanical heat-recovery ventilation system.

Getting planning permission for the build was "a nightmare from the start": the local planners not liking anything about the project – neither the hempcrete nor the plan for the amalgamation of the two houses. Luckily, however, the planning committee saw things differently and granted permission. After some initial problems with local authority building control, Bill and Jo switched to using an Approved Inspector, which smoothed the rest of the process considerably.

The scope of the project they have taken on would be daunting to many, but Bill is clearly a man who is not happy unless he is getting his teeth into a challenge. Bill and Jo designed and detailed the building themselves with help from Westwind Oak, who drew up the plans for the oak frame in the extension, based on Bill's original ideas. The build was completed by Bill and Jo together with Dave, with a little assistance from subcontractors along the way. Once they had worked out how to do it, they found hempcrete very easy to use, and affordable, although they did not make any direct cost comparisons with other materials.

The bulk of the work has been carried out by the family, with Dave working on it while Bill was out at work, and Bill taking over out of office hours. This shows in the level of care and attention to decorative detail visible in the building, and the overall effect is of a home that has been built with love. Although they were often learning the necessary skills as they went along, it is clear from the result that what Bill, Jo and Dave lacked in experience they more than made up for in the care they took with the work, and that they invested a bit of their personalities into every wall built and finish applied.

So, with a kitchen in the hempcrete extension and a few severe winters having passed since the hempcrete solid-wall insulation was installed, are Jo and Bill happy with the thermal performance of the hempcrete? "More than happy!" Jo says: "We've been through three hard winters with just the wood burner in the living room." The underfloor heating, which is fed by a ground source heat pump, has been in for a while now, but Jo says "we haven't used it – we haven't needed to – and we've only bought £150-worth of firewood in that time, to supplement what we already had from some trees we had to cut down in the garden. We've got no heating either, except heated towel rails upstairs fed from a small gas boiler, which also provides domestic hot water – no radiators at all."

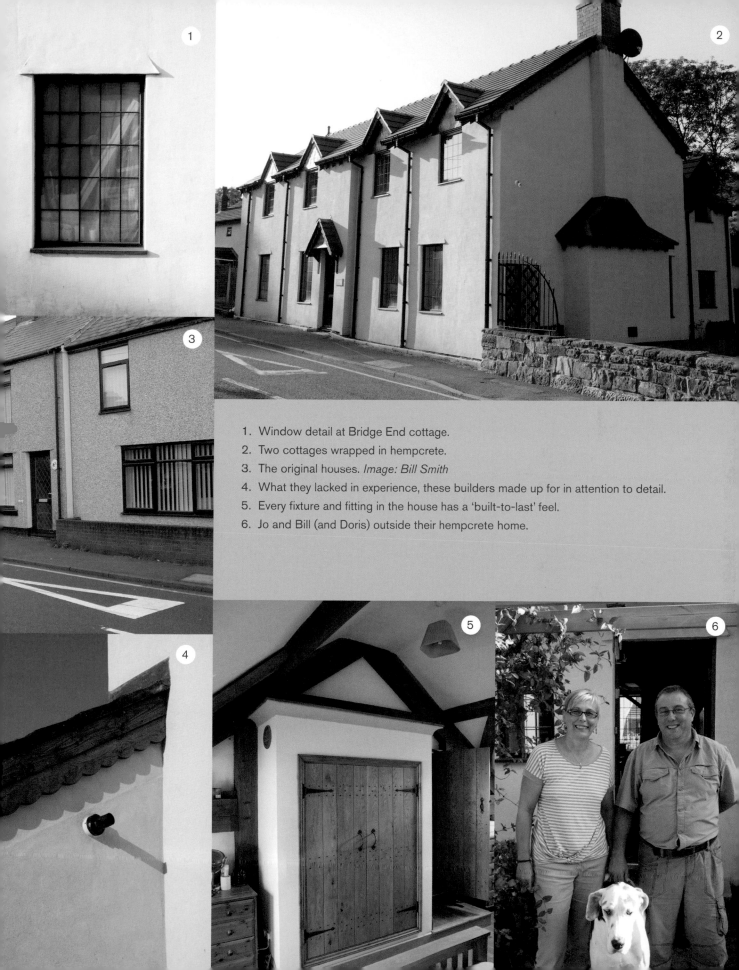

1. Window detail at Bridge End cottage.
2. Two cottages wrapped in hempcrete.
3. The original houses. *Image: Bill Smith*
4. What they lacked in experience, these builders made up for in attention to detail.
5. Every fixture and fitting in the house has a 'built-to-last' feel.
6. Jo and Bill (and Doris) outside their hempcrete home.

CHAPTER TWENTY-THREE

A look to the future

So, what next for hempcrete? In bringing to a close what is intended to be the most comprehensive account to date of this exciting new method of building, it is only natural to ask this question. Perhaps it is an appropriate moment to reflect on the place hempcrete has carved for itself within construction in the UK and beyond.

Over a decade has passed since the first hempcrete houses were built in the UK, and the use of this material in countless projects in that time – from one-off self-builds to housing estates to large-scale commercial and industrial buildings – speaks for itself. The application of hempcrete for upgrading the thermal performance of heritage buildings is almost an industry in itself, and, with retrofit of insulation to older properties becoming an increasingly high priority, we are sure to see the use of hempcrete in this sector continuing to grow. As a reliable breathable insulation material that works perfectly *with* the existing fabric of traditional buildings, it is unrivalled in its versatility and unique thermal performance.

While hand-placed cast-in-situ hempcrete has been the 'core' methodology to date, practices are continuing to evolve, with increasing use of spray application and pre-cast systems, especially for large-scale projects. The fact that hempcrete has been adopted so quickly in such a wide range of contexts, despite being a relatively new and low-profile material, is a strong indication that it will continue to become more widely accepted and employed.

From its original development in France, the use of hempcrete quickly spread internationally, with the highest levels of use seen in European countries, including the UK, and growing numbers of hempcrete buildings going up in Canada, Australia and South Africa. The next big market to open up will surely be the USA, with the planting in 2013 of the first 'legal' hemp crop of a significant size in many decades. At the time of writing, the bizarre legal situation persists in the USA that industrial hemp is legal to import into the country, and use, in its processed state, but *illegal* to grow there, as it comes under Federal

Left: Hempcrete houses with a striking modern design.

The availability of hemp shiv for building is increasing in Europe and across the world.

While not exactly complicated, the skills needed to use hempcrete successfully are not currently widespread in the construction industry.

drug prohibition laws. However, in the face of mounting public pressure, the first states have begun to make industrial hemp legal under State law, so surely it can only be a matter of time before its cultivation is allowed across the United States. To our knowledge, a hempcrete house under construction in Oklahoma at the time of writing is only the fifth of its kind in the USA, but we will surely see rapid change in this area over the next few years.

Given the relatively low-tech construction method, and the advantage of being able to use locally grown crops to provide an environmentally sustainable construction material, we would expect to see a key future market in developing countries. That there has been limited use of hempcrete in the developing world to date is possibly due to supply chain issues, or a general lack of awareness about the material.

The UK market has, until recently, suffered from having a limited number of binder materials and hemp shiv on the market, but this situation is now starting to change, with a choice of three binders now available and, at the time of writing, another three in development. The hemp shiv

market is starting to open up, with independent shiv producers and French hemp processors looking to expand into the UK.

The key challenge for the UK industry now is how to take forward the development of agreed standards of construction across the industry and improve the standardization and testing of binder and hemp shiv materials. The aim should be that those wishing to build with cast-in-situ hempcrete can do so with the confidence that comes from acting within clear best-practice guidelines, with materials which have a high degree of consistency and quality control.

The development of skills and knowledge across the wider construction industry is also vital. The importance of application technique in the use of hempcrete means that there is a risk, when

A self-build barn conversion using hempcrete for external insulation and internal walls.

used with a lack of experience, of poor standards in application causing problems, particularly with regard to extended drying times. This then puts companies off using the material again, especially in large commercial projects. Such 'false starts' are perhaps unsurprising in what is essentially the start of a huge paradigm shift within the construction industry.

Hempcrete is the first natural, sustainable construction material to be adopted on any significant scale in the mainstream UK construction industry. This is no small feat in an industry that is notoriously suspicious of change and of new, 'untried' materials. Hempcrete has on its side the benefit of being, compared with straw bale or cob, for example, a material that has a passing resemblance to things that are understood by the conventional builder. The

materials arrive on-site on pallets and, while the bales of hemp usually raise a few eyebrows, the binder is reassuringly packaged in bags like cement, and the construction method of building temporary shuttering and mixing and casting the material is easily recognizable to the modern builder.

Despite all this, however, hempcrete does *not* act like the conventional materials that the builder of the late twentieth century is used to. Thought and proper understanding of the material are required to apply it successfully; in other words, there is a need for increased understanding across the industry not only of *why* we should use hempcrete, but also of *how* we should use it.

We hope that, in some small way, this book furthers that aim.

Notes*

Chapter 1

1 Roulac, J. W. (1997) *Hemp Horizons: The comeback of the world's most promising plant.* Chelsea Green Publishing: White River Junction, VT.

2 Robinson, R. (1996) *The Great Book of Hemp: The complete guide to the environmental, commercial, and medicinal uses of the world's most extraordinary plant.* Park Street Press: Rochester, VT.

3 Williams, M. (2006) *Bridport and West Bay: The buildings of the flax and hemp industry.* English Heritage: Swindon.

4 Adapted from *Hemp Horizons* by J. Roulac, page 15.

5 Raabe, S. (2013) First major hemp crop in 60 years is planted in southeast Colorado. Article in *The Denver Post.* www.denverpost.com/ci_23232417/first-major-hemp-crop-60-years-is-planted

Chapter 2

1 Energy consumption in cast-in-situ Tradical® Hemcrete® buildings has been noted to drop by as much as 75% in the first three years of occupancy. Patten, M., Hempcrete Projects Ltd. (June 2013). Personal communication.

2 Tŷ-Mawr Ltd. www.lime.org.uk/directlinkdownloads/10226Ty-Mawr_lime_hemp.pdf (page 3).

3 Tŷ-Mawr Ltd. www.lime.org.uk/directlinkdownloads/10226Ty-Mawr_lime_hemp.pdf (pages 2 & 3).

4 Womersley's Ltd. www.womersleys.co.uk/acatalog/Hemp_Lime_Plaster_A%20Guide.pdf (page 4).

5 Woolley, T. (2013) *Low Impact Building: Housing using renewable materials.* Wiley-Blackwell: Chichester (page 11).

6 Black Mountain's literature states that their production processes "use up to 90% less manufacturing energy compared to alternative insulation products". www.blackmountaininsulation.com/NatuHemp_Technical_Sheet.pdf

7 Black Mountain Insulation Ltd. www.blackmountaininsulation.com/NatuHemp_Technical_Sheet.pdf

8 At the time of writing, Black Mountain's literature states that their hemp-fibre insulation quilt contains 95% hemp fibre and 5% recycled adhesive binder (www.blackmountaininsulation.com/NatuHemp_Technical_Sheet.pdf), whereas Thermafleece describe their hemp-fibre insulation as containing 60% hemp, 20% recycled polyester, 10% polyester binder and 10% ammonium polyphosphate (a flame-retardant chemical) (www.thermafleece.com/sites/default/files/downloads/Thermafleece%20Hemp%20MSDS.pdf; page 1).

Chapter 4

1 Centre for Alternative Technology (2010) *Zero Carbon Britain 2030: A new energy strategy.* CAT Publications: Machynlleth.

2 Zero Carbon Hub (2013) *Zero Carbon Strategies for Tomorrow's New Homes.* NHBC Foundation: online only. www.zerocarbonhub.org/sites/default/files/resources/reports/Zero_Carbon_Strategies_for_Tomorrows_New_Homes.pdf (page 4).

3 Woolley, T. (2013) *Low Impact Building: Housing using renewable materials.* Wiley-Blackwell: Chichester (pages 125-6).

4 Hens, H. (2012) *Passive Houses: what may happen when energy efficiency becomes the only paradigm?* Online only. www.thefreelibrary.com/Passive Houses: what may happen when energy efficiency becomes the...-a0295268301

5 Woolley, T. (2013) *Low Impact Building: Housing using renewable materials.* Wiley-Blackwell: Chichester (page 138).

Chapter 5

1 Cited in Woolley, T. (2013) *Low Impact Building: Housing using renewable materials.* Wiley-Blackwell: Chichester (page 138).

2 Lime Technology Ltd. (2007) Tradical® Hemcrete® Information Pack. As cited in Woolley (ibid.).

*These notes are available as a pdf with live hyperlinks at: www.greenbooks.co.uk/hempcrete-notes-and-biblio

Chapter 6

1 Sources of information: Batichanvre®: *Hemp Mortars Using St Astier Natural Hydraulic Limes and Products.* www.stastier.co.uk/nhl/guides/pdfs/Hemp_Mortars.pdf

Prompt Natural Cement: *Le Prompt Vicat: Hemp Solutions. Application guide: concrete and hemp mortars.* www.cornishlime.co.uk/pdfs/Hemp%20and%20Prompt%20Solutions.pdf

Tradical® HB: www.tradical.com and *Tradical® Hemcrete® Installers Information Pack.* www.limetechnology.co.uk/pdfs/Tradical_Hemcrete_Installers_Pack.pdf

Chapter 7

1 Bevan, R. and Woolley, T. (2008) *Hemp Lime Construction: A guide to building with hemp lime composites.* IHS BRE Press: Bracknell (page 75).

2 Daly, P., Ronchetti, P. and Woolley, T. (2013) *Hemp Lime Bio-composite as a Building Material in Irish Construction.* Environmental Protection Agency Ireland: online only. http://erc.epa.ie/safer/iso19115/displayISO19115.jsp?isoID=202 (page 44).

Chapter 13

1 Lancashire, R. and Taylor, L. (2011) *Timber Frame Construction* (5th edition). TRADA Technology Ltd.: High Wycombe.

2 More information can be found in *Framing Roofs*: see www.tauntonstore.com/for-pros-by-pros framing-roofs-071229.html

Chapter 16

1 Lime Technology Ltd. (www.limetechnology.co.uk) Tradical® Hempcrete® (contractor FAQ document).

Chapter 20

1 May, N. and Rye, C. (2012) *Responsible Retrofit of Traditional Buildings: A report on existing research and guidance with recommendations.* SPAB: online only. www.spab.org.uk/downloads/STBA%20RESPONSIBLE-RETROFIT.pdf

2 English Heritage (2011) *Energy Efficiency and Historic Buildings: Application of Part L of the Building Regulations to historic and traditionally constructed buildings.* English Heritage: online only. www.english-heritage.org.uk/publications/energy-efficiency-historic-buildings-ptl/eehb-partl.pdf

3 www.english-heritage.org.uk/professional/advice/hpg/compliantworks/buildingregs/

Chapter 21

1 Daly, P., Ronchetti, P. and Woolley, T. (2013) *Hemp Lime Bio-composite as a Building Material in Irish Construction.* Environmental Protection Agency Ireland: online. http://erc.epa.ie/safer/iso19115/displayISO19115.jsp?isoID=202 (page 48).

Chapter 22

1 Shea, A., Lawrence, M. and Walker, P. (2012) 'Hygrothermal performance of an experimental hemp–lime building'. *Construction and Building Materials* 36, 270-275.

2 Daly, P., Ronchetti, P. and Woolley, T. (2013) *Hemp Lime Bio-composite as a Construction Material in Irish Construction.* Environmental Protection Agency Ireland: online only. http://erc.epa.ie/safer/iso19115/displayISO19115.jsp?isoID=202 (page 57).

3 Woolley, T. (2013) *Low Impact Building: Housing using renewable materials.* Wiley-Blackwell: Chichester (page 114).

Glossary

Text in **bold** refers to terms that appear elsewhere in this glossary.

Airtightness A measurable property: the degree to which air leaks through the **thermal envelope** in a building. Leakage of warm air can account for significant energy wastage, especially in old buildings. The Building Regulations set out standards for levels of airtightness in new buildings. Airtightness in a particular building is tested by artificially raising the air pressure inside and measuring the volume of air leakage, per hour, per m^2 of thermal envelope, at an internal/external air pressure differential of 50 pascals. Thus airtightness is expressed as m^3/h/m^2@50Pa (metres cubed [of air entering/leaving the building] per hour per square metre [of thermal envelope] at a pressure of 50 pascals).

Carbon profiling A method of assessing both **embodied energy** and **energy in use**, at the same time, to allow comparison between different methodologies in the built environment. Carbon profiling is a mathematical process used to calculate how much CO_2 is put into the atmosphere from 1m^2 of a building per year.

This enables very different buildings constructed using different technologies to be subjected to meaningful comparison, which in turn allows governments, investors and owner-occupiers to make informed choices about the most sensible carbon-saving measures on which to spend money.

Embodied energy A measure of the total energy use associated with a material at all stages of its life, or in the delivery of a service, considered as if this associated energy was 'embodied' in the final product itself.

Energy in use The carbon emissions associated with the occupation and use of a building. This includes, for example, the energy used for heating and/or cooling the building and for powering appliances. Many of the energy-saving technologies that have become common-place in the UK construction industry today focus on reducing energy in use, but fail to consider the implications of the **embodied energy** used in the manufacture of their materials.

Future-proof A term used to denote the ability of (for example) a building to withstand the stresses and strains of future events: specifically, in this context, the anticipated decreasing availability of fossil fuels and associated high energy costs, and the demand for energy-efficiency measures within society as a whole. A future-proof building can be seen as 'protecting' the owner or occupier against the uncertainties of our energy future.

Hygroscopicity A property of a substance, referring to its ability to attract and hold moisture from the ambient environment. A hygroscopic building material will take up moisture in response to an increase in the **relative humidity** of its surroundings, releasing it again when the humidity drops.

Lifecycle analysis (LCA) (also known as lifecycle assessment) A technique used to assess environmental impacts associated with all stages of a product's life, including raw material extraction; materials processing; manufacture, distribution, use, repair and maintenance; and disposal, recycling or reuse at end of life: a 'cradle-to-cradle' approach. Lifecycle analysis is seen as the 'gold standard' in assessment of the environmental impact of any product or service; however, the costs associated with such in-depth analysis mean that it remains a rarity, especially in the UK, where (compared with Europe) legislation has stopped short of compelling companies to carry out lifecycle analysis on their products.

Low impact Describes doing anything in such a way as to have minimum impact on the environment. This includes an emphasis on sourcing products locally and favouring natural sources of materials instead of highly processed synthetic materials, which often have harmful by-products or use a lot of energy in their manufacture and distribution. Low-impact buildings are constructed using natural materials with low **embodied energy**, and often involve relatively simple techniques that hark back to traditional ways of building.

Off-gassing The evaporation of chemicals, especially **volatile organic compounds (VOCs)** from a material. Many modern mainstream construction materials, including coatings such as paint or varnish, glued products such as composite boards, and insulation products such as those synthetic insulations that contain formaldehyde, give off potentially harmful chemicals at atmospheric pressure and room temperature.

Relative humidity The amount of water held, as vapour, in a body of air at a given temperature (the capacity of air to hold water increases as its temperature increases). Relative humidity is expressed as a percentage of the total maximum amount of water the body of air *could* hold, at saturation point, at that temperature.

A relative humidity in the range of 40-60 per cent is most conducive to human health. There is an increased risk of respiratory tract conditions for those living in an environment that is too dry or too wet.

Thermal bridge (also known as **cold bridge**) A break in an insulation layer that allows heat to bypass it. This is usually found where a material with high thermal conductivity interrupts or penetrates the insulation layer.

Thermal conductivity A measure of a material's ability to conduct heat. Often denoted as the 'lamda value' or 'K-value', thermal conductivity is measured in W/mK (watts per metre kelvin). The thermal conductivity of a material, unlike its **U-value**, is always the same irrespective of its thickness, since it is a constant characteristic of the material itself.

Thermal envelope The area of floors, walls, windows and roof or ceiling that contains the building's internal warm heated volume.

Thermal mass The ability of a body of material within a building to absorb, store and subsequently release heat, as a function of both its specific heat capacity (the amount of heat energy required to raise the temperature of 1 kilogram of a material by 1°K) and its mass. This is defined as the number of joules required to raise the temperature of the entire body of material by 1°K.

A body of material of a given size will therefore have a higher thermal mass if it has a higher density, since its mass will be greater. Buildings built with high-density materials such as stone, brick or concrete have a high thermal mass and therefore respond slowly to changes in temperature, absorbing and storing heat during the day and releasing it slowly through the night.

Thermal performance A material's success, or otherwise, in conserving heat and power in a building. The overall thermal performance of a material used in a given situation may result from different properties of the material, e.g. its insulation value and/or its thermal mass.

U-value A measure of how easy it is for heat to pass through a material, or build-up of materials, therefore describing how good an insulator the material is. U-values are measured in W/m²K (watts per square metre per degree kelvin), and are routinely used in construction to describe how much heat loss there is through the a wall, roof or floor that forms part of the building's **thermal envelope**. The lower the U-value, the better the material performs as an insulator.

Vapour permeable ('breathable') The degree to which a material allows the passage of water vapour (but not necessarily liquid water) through it. 'Breathable' is a (slightly misleading) term in common usage in the construction industry to describe a material with good vapour permeability. It is desirable to maintain vapour permeability when building with natural materials, as the free passage of moisture prevents the formation of damp, which can harm the building's fabric as well as its occupants.

Volatile organic compounds (VOCs) Organic compounds, often used as solvents in construction materials, that evaporate very readily. Certain VOCs may be air pollutants in their own right, but they also cause chemical or photochemical reactions in the atmosphere that cause damage to vegetation, materials and human health.

Zero carbon Refers to an ambitious target set by the UK government that all new homes should be 'zero carbon' by the year 2016. The concept of 'zero carbon' relates to the development of methods of construction, coupled with energy-conservation technologies in the building, that result in zero-carbon emissions or a net carbon saving during the life of the building. This target has now been watered down beyond all recognition, probably due in no small part to the sheer impossibility of achieving it within such a short space of time in a sector that is currently responsible for more than half of the UK's total carbon emissions.

Resources

Australia.....................................356
Canada 356
New Zealand 357
UK 354
USA 356

Not all of the organizations listed below have worked with the authors; inclusion in the list does not imply a personal recommendation.

UK

General information

Alliance for Sustainable Building Products (ASBP)
www.asbp.org.uk
Promotes building products that meet demonstrably high standards of sustainability.

Association for Environment Conscious Building (AECB)
www.aecb.net
Promotes sustainable building.

BRE Group (formerly Building Research Establishment)
www.bre.co.uk
A research-based consultancy; a testing and training organization.

Building Limes Forum
www.buildinglimesforum.org.uk
Encourages the appropriate use of building limes.

Low-Impact Living Initiative (LILI)
www.lowimpact.org
Promotes global sustainable, low-impact ways of living, including low-impact building.

Natureplus
www.natureplus.org
An international organization that aims to increase sustainability within the construction sector through the development of a meaningful standards mark for sustainable building products.

Sustainable Traditional Buildings Alliance (STBA)
www.stbauk.org
A forum for developing the sustainability of traditional buildings.

Timber Research and Development Association (TRADA)
www.trada.co.uk
Internationally recognized as a centre of excellence in the specification and use of timber and wood products.

Hempcrete services

Contractors specializing in cast-in-situ hempcrete

Hemp-LimeConstruct
www.ukhempcrete.com

SHALCO
www.shalco.uk.com

Contractors specializing in pre-cast hempcrete panels

Greencore Construction
www.greencoreconstruction.co.uk

A selection of architects with hempcrete experience

Chris Davies Architect
Gloucestershire
www.chrisdaviesarchitect.co.uk

Ecological Architecture
Inverclyde/Perthshire
www.ecological-architecture.co.uk

Glenn Howells Architects
London/Birmingham
www.glennhowells.co.uk

Justin Smith Architects
Derby
www.justinsmitharchitects.co.uk

Mark Hines Architects
London
www.markhines.co.uk

Modece Architects
Suffolk
www.modece.com

Native Architects
Yorkshire
www.nativearchitecture.co.uk

Noel Wright Architects
Hampshire
www.noelwrightarchitects.co.uk

Rachel Bevan Architects
Co. Down
www.bevanarchitects.com

Vincent and Gorbing Chartered Architects and Town Planners
Hertfordshire
www.vincent-gorbing.co.uk

Specialist hempcrete design and consultancy

Hemp-LimeConstruct
www.ukhempcrete.com

Native Architects
Yorkshire
www.nativearchitecture.co.uk

Professor Tom Woolley
c/o Rachel Bevan Architects
(see above)

Mixer hire

Kilworth Machinery
www.kilworthmachinery.co.uk
Forced-action pan mixers.

Multi-Marque Production
Engineering
www.multi-marque.co.uk
Forced-action pan mixers and roller
pan mixers suitable for preparation
of lime putty mortars.

Training

Building Limes Forum
www.buildinglimesforum.org.uk
Their website lists organizations that
offer training in the use of building
limes.

Centre for Alternative
Technology
www.cat.org.uk
Short courses and postgraduate-level
education. Visitor centre demonstrating practical solutions.

Hemp-LimeConstruct
www.ukhempcrete.com
Offers training in building with
hempcrete: both a basic training
course and bespoke on-site training
for organizations.

Low-Impact Living Initiative (LILI)
www.lowimpact.org
Offers training in a range of
sustainable construction techniques,
including hempcrete.

Hempcrete material Suppliers

Binders

Anglia Lime
www.anglialime.com
Also offers aggregates and limewash

Cornish Lime Company
www.cornishlime.co.uk

Hemp-LimeConstruct
www.ukhempcrete.com

Lime Green
www.lime-green.co.uk

Womersley's
www.womersleys.co.uk

Hemp shiv

Hemp-LimeConstruct
www.ukhempcrete.com

K J Voase and Son
www.kjvoaseandson.co.uk

Lime Green
www.lime-green.co.uk

Tŷ-Mawr
www.lime.org.uk

Hempcrete blocks

H. G. Matthews Traditional Brick
Makers
www.hgmatthews.com

Hemp-LimeConstruct
www.ukhempcrete.com

Hempcrete panels

Greencore Construction
www.greencoreconstruction.co.uk
www.limetechnology.co.uk

Other natural and traditional building materials

Back to Earth
www.backtoearth.co.uk
Clay boards and plasters, wood-fibre
insulation boards, lime-based render
systems.

Black Mountain
www.blackmountaininsulation.com
Hemp, flax and sheep's-wool quilt
insulation.

Clayworks
www.clay-works.com
Clay plasters, books and training on
clay plastering.

Cornish Lime Company
www.cornishlime.co.uk
Lime products, aggregates, limewash, breathable paints, lime–hemp
plasters.

Ecomerchant
www.ecomerchant.co.uk
Natural insulations, lime plasters,
wood wool boards.

Lime Green
www.lime-green.co.uk
Wood wool boards, lime products,
breathable paints, lime–hemp
plasters and hemp shiv.

Lime Stuff
www.limestuff.co.uk
Lime products, limewash,
aggregates, breathable paints,
decorating materials.

Mike Wye and Associates
www.mikewye.co.uk
Lime products, limewash, and
breathable paints, related training on
how to apply them, traditional tools
and materials.

Natural Building Technologies
www.natural-building.co.uk
Wood-fibre insulation boards,
lime-based render systems.

Natural Insulations
www.naturalinsulations.co.uk
Sheep's-wool quilt insulation,
wood-fibre insulation boards.

Rose of Jericho
www.rose-of-jericho.demon.co.uk
Traditional lime plasters, limewash,
traditional paints.

Thermafleece
www.thermafleece.com
Hemp and sheep's-wool quilt insulation.

Tŷ-Mawr
www.lime.org.uk
Wood-fibre insulation boards, wood wool boards, lime products and plassters, aggregates, limewash, breathable paints, limecrete.

USA

Alembic Studio
http://alembicstudio.com/
Hempcrete home design services.

American Hemp
www.americanhempllc.com
Suppliers of hemp shiv for building.

American Lime Technology
www.americanlimetechnology.com
Suppliers of hemp building materials, hemp fibre insulation and lime renders and mortars.

Cloverdale Equipment, LLC.
www.cloverdaleequip.com
Mixer hire company whose mixers have been used for mixing hempcrete by US self-builders.

Hemp Solutions
http://hemp-solutions.org/
Supply of hemp hurd for building.

Hempsteads
http://hempsteads.info/
Hempcrete information and consultancy from hemp builder-designer Tim Callaghan.

Hemp Technologies Global
www.hemp-technologies.com
Suppliers of hempcrete materials and hemp-fibre insulation. A source of information on many uses of the hemp plant including building. Initial training is provided for those purchasing their hempcrete building materials.

Hemp Traders
http://www.hemptraders.com/default.asp
Supply of hemp hurd for building.

National Hemp Association
http://nationalhempassociation.org/
Promoting the hemp industry in the United States.

North American Industrial Hemp Council
http://www.naihc.org/
Campaigning and educational organization.

North Carolina Industrial Hemp Association
http://www.ncindhemp.org/

Original Green Distribution
www.originalgreendistribution.com
Suppliers of hempcrete building materials and hemp-oil wood preservatives.

Tiny Hemp Houses
www.tinyhemphouses.com/
Hempcrete contractor services.

Trans Mineral USA
http://www.limes.us/
Suppliers of hempcrete binder (Batichanvre) as well as natural hydraulic limes and putty limes for plastering, and lime paints.

Vote Hemp
http://www.votehemp.com/index.html
Lobbying organisation promoting legalisation of industrial hemp across all US states.

Canada

Canadian Hemp Trade Alliance
www.hemptrade.ca
Organization promoting Canadian hemp and hemp products globally.

The Endeavour Centre
www.endeavourcentre.org
Offers training in the use of hempcrete and other natural building materials.

Hempcrete Natural Building Ltd.
www.hempcrete.ca
Offers hempcrete training and consultancy.

North American Hemp and Grain Co. Canada
www.hempcanadabulk.com
Provides hemp for building, although this is not specifically highlighted on their website.

The Lime Plaster Co.
http://www.naturallimeplaster.ca/
Specialist plastering company offering lime and clay plastering in traditional and modern buildings.

Plains Hemp
http://plainshemp.com/
Suppliers of hemp shiv for building.

Australia

Australian Hemp Foundation
www.australianhemp.org
Industrial hemp industry organization.

Australian Hemp Masonry Company
www.hempmasonry.com
Suppliers of hemp shiv, hempcrete binder and lime renders. Also provides mixer hire, hempcrete training and hemp building consultancy.

Australian Hemp Party
http://australianhempparty.com
Campaigning organization promoting the legal cultivation of hemp across Australian states.

Developing Sustainable
Directions (Green Building
Australia website)
www.greenbuildingaustralia.com.au
Organization promoting sustainable
construction methods including
hempcrete and hemp production
generally.

Ecofibre industries
www.ecofibre.com.au
Suppliers of hemp fibre and shiv for
a range of products/industries. Not
specifically marketing shiv for
hempcrete building on their website.

Hempcrete Australia
www.hempcrete.com.au
Suppliers of hempcrete materials,
hempcrete mixers, and certified
training for hempcrete installers in
both Australia and New Zealand.

Hemp Farm Australia
www.hempfarm.com.au
Hemp Company offering advice on
growing and all uses of hemp,
including building.

Industrial Hemp Association of
New South Wales
www.ihansw.org.au
Organization promoting the
upscaling of industrial hemp
production in New South Wales.

Industrial Hemp Association of
Queensland
www.ihaq.com.au
Non-profit organization promoting
growing and use of hemp.

Industrial Hemp Association of
Victoria
www.hempvictoria.org
Non-profit organization promoting
growing and use of hemp.

Northern Rivers Hemp Association
www.northernrivershemp.org
Supply of hemp shiv, hempcrete and
related materials, and hempcrete
training. Promoting the hemp
industry generally.

Oz Hemp
www.ozhemp.com.au
Hempcrete material supplier.

Sowden Building Solutions
**www.sowdenbuildingsolution.com.
au**
Sydney building contractors with
experience of hempcrete.

New Zealand

Hemp Technologies Global
www.hemp-technologies.com/nz
Suppliers of hempcrete materials,
and training in building with
hempcrete. Promoting hemp
generally for a range of uses.

Purity Homes
www.purityhomes.co.nz
Suppliers of hempcrete products
together with a range of other
natural building products.

New Zealand Hemp Industries
Association
http://www.nzhia.com/
Campaigning and professional
organization for industrial hemp in
New Zealand.

Bibliography*

Hempcrete

Arnaud, L. and Amziane, S. (2013) *Bio-aggregate-based Building Materials: Applications to Hemp Concretes*. Wiley-ISTE: London.

Bevan, R. and Woolley, T. (2008) *Hemp Lime Construction: A guide to building with hemp lime composites*. IHS BRE Press: Bracknell.

Daly, P., Ronchetti, P. and Woolley, T. (2013) *Hemp Lime Bio-composite as a Building Material in Irish Construction*. Environmental Protection Agency Ireland: online only. http://erc.epa.ie/safer/iso19115/displayISO19115.jsp?isoID=202

Woolley, T. (2006) *Natural Building: A guide to materials and techniques*. The Crowood Press: Marlborough.

Lime

Cowper, A. D. (1927) *Lime and Lime Mortars*. HMSO. A special report for the Building Research Station (now BRE). Facsimile Edition published (1998) by Donhead Publishing: Shaftesbury.

Henry, A. and Stewart, J. (2011) *Practical Building Conservation: Mortars, renders and plasters*. Ashgate Publishing: Farnham.

Holmes, S. and Wingate, M. (2003) *Building with Lime: A practical introduction*. Practical Action Publishing: Rugby.

Schofield, J. (1997) *Lime in Building: A practical guide*. Black Dog Press: London.

Weismann, A. and Bryce, K. (2008) *Using Natural Finishes: A step-by-step guide*. Green Books: Cambridge.

Natural paints and coatings

Edwards, L. and Lawless, J. (2002) *The Natural Paint Book: A complete guide to natural paints, recipes, and finishes*. Rodale Books: Emmaus, PA.

Woolley, T. (2006) Chapter 9: Paints and Finishes; in *Natural Building: A guide to materials and techniques*. The Crowood Press: Marlborough.

Woolley, T., Kimmins, S., Harrison, P. and Harrison, R. (1997) Chapter 9: Timber Preservatives; Chapter 11: Paints and Stains for Joinery; in *The Green Building Handbook Vol 1*. Spon Press: London.

www.ecodecorators.org/products.php

www.theguardian.com/lifeandstyle/2009/feb/09/eco-natural-paints-guide-best

www.lowimpact.org/factsheet_natural_paints.htm

Retrofitting traditional buildings

English Heritage (2011) *Energy Efficiency and Historic Buildings: Application of Part L of the Building Regulations to historic and traditionally constructed buildings*. English Heritage: online only. www.english-heritage.org.uk/publications/energy-efficiency-historic-buildings-ptl/eehb-partl.pdf

May, N. and Rye, C. (2012) *Responsible Retrofit of Traditional Buildings: A report on existing research and guidance with recommendations*. SPAB: online only. www.spab.org.uk/downloads/STBA%20RESPONSIBLE-RETROFIT.pdf

Suhr, M. and Hunt, R. (2013) *The Old House Eco Handbook: A practical guide to retrofitting for energy-efficiency & sustainability*. Frances Lincoln: London.

Timber framing

Benson, T. (1999) *Timberframe: The art and craft of the post-and-beam home*. The Taunton Press: Newtown, CT.

Chappell, S. (2005) *A Timber Framer's Workshop*. Fox Maple Press: Brownfield, ME.

Lancashire, R. and Taylor, L. (2011) *Timber Frame Construction* (5th edition). TRADA Technology Ltd.: High Wycombe.

Law, B. (2010) *Roundwood Timber Framing: Building naturally using local resources*. Permanent Publications: East Meon.

Index

Page numbers in *italic* refer to figures and illustrations.

access equipment 107, 261
   see also scaffolding
acoustic performance 33, 88, 89, 91, 97-8,
   99, 126, 131, 133, 225, 311, 330
adhesives 55, 268, 280
Adnams' distribution warehouse *32*, 33
Agan Chy 124-7, *125*, *127*
aggregates 151, 152, 222, 224, 233, 234, 237,
   240, 243, 244, *313*, *316*, 318, *318*, 335
   plant-based 23, 25, 77, 88, 89, 94, 98,
      133, 134, 140, 192, 209, 290, 293, 296
air limes 42-4, 45, 46, 50, 51, 80-1, 93-4,
   215, 236, 263
airtightness 35, 36, 39, 55, 65, 81, 97, 126,
   143, 250, 296, 297, 299, 309-10, 331
   air leakage 55, 65, 160, 270, 271, 274,
      280, 296, 310, 311
   air quality and 311
   airtight line 310, 311, 323
   at junctions 39, 297-9, *297*, 311-12, 317,
      325, 326, *326*
   cladding *297*, 299, 311, 331, 333
   in conventional construction 65
   hempcrete airtightness test 312
   roofs 322, 323, 324-6, 327, 328, 329
   thermal envelope 120, 221, 310, *310*
alcoves and cupboards 136, 137, 159, 186
allergies 65
alumina 45
angle brackets 161, *161*, *162*, *163*, 300, 322
Anutone 183
Approved Inspectors 82, 92, 126, 149, 344
architects 120, 131, 133-4, 147, 290, 292,
   309, 341, 354-5

barn conversion *349*
barrier creams 112, 210-11
bast fibres 15, 19, 21, 25, 26, 37, 38, 233
   see also fines
bathrooms, finishes 249-50, 284
Batichanvre 80, 82, 223
Baumit 294, 335
bell mixers *100*, 105, 106, 201-5, *204*
binders 23, 24-26, 51, 71, 74, *81*
   characteristics 81, 82
   faulty binders 75-6, *75*, *76*
   functions 80
   heritage building application 273-4

hygroscopicity 93-4
   lime-based *see* lime-based binder
   mixing your own 75, 81, 83
   open-source binder recipe 83
   polymer binder 56
   proprietary products 51, 75, 80, 81-4
   researching 120
   shrinkage 153, 274, 296
   storage 258
   suppliers 289, 355
   temperature parameters 123, 263
   UK pre-approval testing 84
   vapour permeability 80, 93-4
   variations in 79-83
bio-aggregate building materials 87
bit holders 105
block, pulley and rope 106
blocks 30, 31-3, *31*, 34, 35, *35*, 36, 51, 68,
   69, 83, 90, 91, *91*, 93, 132, *356*
   bedding mortar 32
   compressive strength 32, 90
   cutting 31
   disadvantages 32
   laying 31, *31*, 32
   reliability and consistency 35
   'structural' blocks 32-3
breathability *see* vapour permeability
bricklayers 255
brickwork
   brick cladding 255, 335
   historic 277
   plinths 151, *152*, *317*
   pointing 48
Bridge End Cottage 342-5, *343*, *345*
Bridport, Dorset 17
brooms and shovels 106
buckets and tubs 105, 106, *106*, 196, 254
   measuring using buckets 202
build practicalities 253-4
   drying management 122, 217, 219,
      264-5
   materials supply 255-6
   planning the build 117-23
   roles
      ferrying team 254, 256
      finishing teams 255
      framing team 253-4
      mixing team 254, 255-6
      placing team 209-10, 214, 254, 256-7
      shuttering team 255, 257, *257*
      switching 255-8

site organization 258-61
   areas for breaks and meals 261
   areas for cleaning up 261
   cutting area 259
   mixing area 259
   scaffolding and access equipment
      260-1
   storage of materials 258-9
   tools and equipment storage 259
   smaller builds 258
   Stage One 253-4
   Stage Two 254-8
   Stage Three 255
   weather 261-4
building control officers 91, 92, 147, 148,
   149, 157, 319-20
Building Limes Forum 43, 354
Building Regulations (UK) 63, 91, 140,
   271, 291, 292, 319
burnt sand mastic 245, *245*, 246, 295, 298,
   312, 317, *337*, *339*

cables and wiring 144, *144*, *145*
calcium carbonate ($CaCO_3$) 41, 42
calcium hydroxide ($Ca(OH)_2$) 42
calcium lime *see* air limes
calcium oxide (CaO) 42
calcium sulphate ($CaSO_4$) 48
Callowlands, Watford *320*
cannabis 15, 17-18
Cannabis sativa see hemp plant
carbon dioxide (CO_2) 7, 9, 38, 42, 48, 53,
   57, 69
carbon profiling 51, 55
carbon sequestration 7, 9, 38, 56, 57, 69, 97
'carbon sink' 57, 69, 120, 222, 325
carbonation 42, 43, 44, 45, 47, 48, 50, 81,
   89, 93, 215, 235, 236, 263
carrier boards 134, 135, 138, 228, 233,
   236, 296, 321, *328*, 330, *332*
   see also magnesium silicate board;
      wood wool board
cast-in-situ hempcrete 22, 27-9, *28*, 51, 64,
   64, 68, 70, 102, 347
   costs 36
   drying *see* drying hempcrete
   embodied energy 36
   placing *see* placing hempcrete
   and pre-cast hempcrete compared 32,
      35-6
   trial casts 73

ceilings
  cast hempcrete 225-7, *225*, *226*, *227*,
    330, *330*
  design detailing 330
  insulation 225-7, 330, *330*
  joists *see* joists
  vapour permeability 225-6
cellulose 23, 56, 63, 94, 218, 314, 328
cement
  mortars 43, 152, 192
  natural cements 23, 45-6, *46*, 47, 50, 51,
    80, 151, 152, 215
  render 58, 62, 72, 142, 221, 244, 255,
    267, 277, *317*
  see also Portland cement; Prompt
    Natural Cement
Chanvrière de l'Aube 27
chipboard 169-70
cladding 72, *141*, 216, 218, 250-1, *250*,
  255, 291, 294, 311, 326, *329*, 330-6,
  334, *335*
  airtightness *297*, 299, 311, 331, 333
  brick 255, 335
  cladding battens 331-2, *332*
  design detailing 330-6
  hung tiles 250, 273, 274, 278, *308*, *333*
  lime render–timber cladding
    combination *250*, 251
  rain-screen 124, 218, 251, 294, 332
  render coating under 331, 332, *332*
  shiplap 251, 332, *332*
  slate 135, 250, 274, 278, 305, 306, 331,
    336
  stone 255, 335
  timber *125*, *141*, 250, *250*, 251, 255,
    306, 331, 332, *332*, *333*
  vapour permeability 250, 330, 331, 332,
    332, 334
  vented air gaps 122, 135, *141*, 250, 291,
    299, 306, 330, 331, *332*, 334, *334*
  with no vented cavity 335-6
clay
  paints 248
  plasters 135, *141*, 234-5, *235*, 249, *278*
coated expanded clay aggregate 222, 224,
  313, *316*, 318, *318*
cob 64, 67, 270, *270*, 349
cold bridging 31, 32, 34, 35, 36, 64, 97,
  131, 153, 222, 228, 327, 328, 329, 336
collapse 26, 69, 76, 77, *77*, 80, 192, 193, 222
compaction 69, 95, 98, 99, *150*, 193-4,
  207-8, 209, 211, 214, 226, 274
  over-compaction 71, 207, 209, 211, 254
compressive strength 32, 90, 215, 313
concrete
  foundations 136, 147-8, 149, *150*, 151,
    301, *314*

frames 33, 132, 155
  lime–pozzolan concrete 151
  plinths 151-2, 153
condensation *60*, 64, 94, 97, 99, 132, 279
construction, hemp in 23-39
  history of 24-5
  large-scale commercial builds 29, 30,
    33, 34, 122
  other uses 37
  see also cast-in-situ hempcrete; pre-cast
    hempcrete
construction industry
  carbon emissions 53, 55-6
  industry standards for hempcrete
    construction, lack of 25, 27, 76, 77,
    83, 88, 99, 289-90, 348
  resistance to new practices 25, 349
  UK 25, 53, 264, 349
construction materials, sustainable 55-7
cordless drills 104, 105
corners
  placing hempcrete 30, 140, 201, 211,
    216, 303, 338
  plastering 238, 243, 244, *244*
  reinforcement 244
  shuttering 177-80, *178*, *179*, 185, 257
costs 10, 30, 35, 71, 73, 119-20, 121, 122,
  123, 149, 155, 170, 185, 197, 216, 255,
  305, 319, 321, 331, 333
  comparisons with conventional
    construction 119, 133
  labour 35, 184, 209, 291-2, 327, 333
  pre-cast hempcrete 36
  shiv 27
  tools and equipment 101-2, 103
  transportation 27, 36, 51, 152, 225
cracking 37, 62, 90, 152, 236, 242, 244,
  263, 290, 291, 321
cuts (wounds) 113, 114

damp-proof course (DPC) *61*, 62, 136,
  151, 152-3, 160, 273, 281, 315, *315*,
  316, 318
damp-proof membranes (DPMs) 58, 135,
  150, 153, 221, 222, 281, 318
day joints *212*
deformability 68, 90, 91, 137, 290
dehumidifiers 106, 265
dermatitis 114
design fundamentals 120, 289-307
  construction process: common issues
    302-4, 340
  awkward placing 303
  eaves, placing hempcrete under
    302-3, *302*
  horizontal timbers 303-4, *303*
  detailing decisions 133, 135, 309-41

essential issues 120, 341
  frame design 138, 155, 157-8, 166, 173
  nature of the material 292-301, 340
    alkaline environment 299-301
    shrinkage *see* shrinkage
    vapour permeability *see* vapour
      permeability
    vulnerability to moisture ingress
      293-6
  planning the build 117-23, *120*
  site topography and climate 304-6, 340
  structural limits of the material 290-2,
    340
    coverage over structural elements
      192, 290-1
    racking strength *see* racking strength
    wall thickness 291-2
diagonal bracing 92, 158, 164-5, 321, 322,
  322
disc grinders 105
door openings 139-40, 295, 303, 317, 336
  door framework 162, 164
  lintels *see* lintels
  placing hempcrete 211, 272, 280
  shuttering 185-6, *186*
drainage 281, 319
drilling into walls 143, 144
drip details 246, *246*, 294-5, *294*, *295*,
  334, 340
dry hydrate of lime 42-3, 47
  see also air limes
drying hempcrete
  dehumidifiers 106, 265
  drying times 70-1, 79, 121-3, 209,
    215-19, 249, 264, 336
    'normal' 217
    wall thickness and 218
  electronic moisture readings 218-19,
    218
  embodied energy 36
  full set 79, 80
  good drying management 122, 217,
    219, 223, 224, 264-5
  hydraulic set 44, 45, 46, 47, 50, 51, 80
  initial set 70, 79, 80, 139, 180, 211, 215,
    264
  off-site 30, 32, 34
  problems arising from 73
  resting moisture content 71, 193, 215
  shrinkage 37, 81, 143, 153, 245, 274,
    296-9, 312, 317
  weather conditions 35, 121, 123,
    217-18, 307
dust 26, 27, 76-7, 79, 98, 110, *110*, 111,
  112, 113, 199

Earthborn 248

eaves 323, *324*, *325*, *326*
  airtightness 323, 326, *326*
  placing hempcrete 302-3, *302*, 323
electrical installations 144, *144*, 145, 280, 300, 303
electronic moisture meters 218-19, *218*
embodied energy 7, 9, 11, 27, 35, 36, 51, 54, 69, 91, 97, 133, 136, 147, 149, 151, 152, 225, 245, 247, 295, 297
  assessing 55
  cast-in-situ hempcrete 36
  conventional construction materials 10, 119-20
  limes 48, 68, 69
  pre-cast hempcrete 32, 34, 35, 36
  shuttering boards 169, 183, 184
  synthetic insulations 67, 221, 331
energy-in-use target 54
English Heritage 268, 269, 271
environmetal impacts of materials and processes 7, 10, 36, 183, 184, 297, 304, 331
expandable tapes 245, 297-8
extensions 157, 297, 320
eye stations 113, *113*
eyes, hazards 110, 114, 191

failure *see* collapse
fat lime *see* air limes
faulty materials 75-7
ferric oxide 45
fibreboard, hemp 37
fines 21, 26, 27, 76, 79, 233
finishes 102, 135, 141-2, 231-51
  advice, sources of 234, 236
  applying too early 36, 74-5, 216, *216*, 264
  costs 71
  finishing teams 255
  floors 224, 251
  for bathrooms and kitchens 249-50
  for permanent shuttering 233-4
  heritage buildings 272, 274, *280*
  judging readiness for 217, 218-19
  literature on 231-2
  site-specific effects on 306
  staining 74-5, *74*, 216, *216*, 217, 218
  vapour permeability 74, 193, 216, 224
  wet-applied finishes 232-49
  see also cladding; limewash; paints; plasters; renders
fire regulations 91
fire resistance 7, 92-3
fire retardants 26, 55, 56
first aid 105, 110, 113-14, 191
fixing into hempcrete walls 137, *138*, 159
flagstones 251, 319

flat roofs 165, 225, 326, 329, *329*
  airtightness 326
floats 107, 233, 239
floors
  conventional construction 221
  damp problems 317
  design detailing 317-19, *318*
  finishes 224, 251, 319
  flood-prone areas 318
  floor slabs 90, 91, 142-3, *220*, 221, 222-3, *223*, 251, *281*, 326
  hempcrete floor insulation 221-4
    hempcrete density 222, 319
    insulating sub-base layer 222, 223, 224
    thickness 222, 319
  joists *see* joists
  levelling 224
  suspended floors 39, 143, 271, 281, *281*, *315*
  synthetic insulations 221
  timber floors 39, 153, 224, 251, 271, *315*, 319
  underfloor heating 224, 251, *316*, *318*, 319, 344
  vapour permeability 221, 222, 317-19
formulated hydraulic limes 45, 47
formwork *see* shuttering
foundations 135-6, 147-51
  building control inspector approval 149
  concrete 136, 147-8, 149, *150*, 151, *301*, *314*
  free-draining 62, 136, 149
  function 147
  limecrete-filled trenches 149
  raft foundations 136, 148, *148*, *150*, 151, *301*, *314*
  rock-filled gabions 149, *149*
  strip foundations 147-8, 149
frame seal beads 245-6, *245*, 295, 338
frames, structural 27-8, 30, 68, *68*, 72, 91-2, 102, 136-8, 155-67
  accurate construction, importance of 102, 167
  central frame 134, 136, 137, 158, *159*, 167, 213
    shuttering 173-7, *174*, *176*
  concrete 33, 132, 155
  double frames 160, 181, 182, 185, 320, 332-3, *332*
  exposed frame 136, 137, 158-9, *159*, 213, 275, *276*, 331
    shuttering 180, 181-3, *182*
  fixings, alkali-resistant 136, 158
  frame design 138, 155, 157-8, 166, 173
  framing team 253-4
  glulam 33, 132, *134*, 136, 155, 156, *156*, *157*, 315, 320

hardwood 135, 138, 156, *156*, 159, 291, 315, 331, 333
  horizontal battens, addition of 159, 213
  infill panels (heritage buildings) 272-3, *273*, 274-5, *276*
  large buildings 155, 156, *156*, *157*
  noggins 91, *162*, 164, *165*, 303, *303*, *314*, 327, *328*, *337*, *339*
  for openings 162, 164
  practical considerations 166-7
  prefabrication in sections 166
  racking strength *see* racking strength
  smaller buildings 157-8
  softwood 28, 134-5, *134*, 136, 137, 138, 155, 157, *159*, *165*, 170, *180*, 315, 321, 331, *332*, *333*, 341
  steel 33, 132, 136, 155, 156, *156*, *157*, 301, *301*, 320
  tools 103-4, 105
France 25, 27, 45, 51, 80, 84, 87, 88, 348
freezing temperatures 262-3
French drains 281, 319
full set 79, 80

gabions 149, *149*, 152
galvanized steel 160, 299, 300
gas supply pipes 142-3
gas-powered heaters 106, 265
GE Sensing Protimeter *218*, 219
geotechnical engineers 149
geotextile membranes 318, *318*, 319
glass-fibre mesh 236, 242, 243-4, *243*, 244, 245, 246, 295, 344
gloves *108*, 111, 112, 196, 210
glulam
  beams *132*, 158, *165*
  frames 33, 132, *134*, 136, 155, 156, *156*, *157*, 315, 320
goggles 107, 111, *111*, 112, 196
Gorilla Tubs 106, 196, 254
grid floats 107, *107*, 188
guttering 294, *294*
gypsum
  plasterboard 49, 133, 134, 183
  plasters 48-9, *49*, 58, 72, 142, 232, 244, 278, 293

hardwood
  beads 245, *245*, 295, *337*
  frames 135, 138, 156, *156*, 159, 291, 315, 331, 333
health and safety 109-15
  first aid 105, 110, 113-14, 191
  hazards 109, 110, 111, 260
  long-term health risks 114
  personal protective equipment (PPE) 109, 111-12, 191, 210

precautionary measures 110-11
safe working practices 109, 112-13,
    191, 260
wet weather precautions 113
health issues 7, 9, 42, 48, 49, 56, 69, 98,
    99, 119, 131, 135, 191, 248, 249, 267,
    268, 327
heaters 106, 107, 265
Hembuild® 33
Hemclad® 33-4
Hemcore *see* Hemp Technology Ltd
Hemcrete Projects Ltd 33
Hemp Lime House 282-5, *283, 285*
hemp plant *14, 26,* 68
    bast fibres 15, 19, 21, 25, 26, 37, 38, 233
    British hemp industry 16-17, 18, 19-21,
        20
    etymology of 15
    growth habit 15, 20, 21
    harvesting 21
    history 15-21
    processing 21, 26-7
    products and uses 7, 16, 19, 25
    psychoactive effect 17-18
    retting 21
    seed crops 21
    shiv *see* shiv
    THC content 18
    US hemp industry 18-19, 347-8
    weed and pest resistance 20
hemp-fibre quilt insulation 25, 33, 38-9,
    39, 274, 276, 296, 297, *297,* 298, *298,*
    317, 327
    at junctions 297, *297, 298, 298*
    hygroscopicity 38
    polyester component 39
    roofs 323, *324, 325,* 326
    window openings *337, 339*
    wrapping pipework in 143
hemp–lime mix 28, 79-85, 140
    density of mix
        consistency 209-10
        higher-density mix 85, 90, 91, 208,
            208, 222, 330
        lower-density mix 85, 90, 91, 96, 137,
            192, 207, 208, *208,* 214, 228, 274
        ratios 70, 84-5, 95, 98, 203, 205, 208
        specific applications 90, 208.84-5
        'standard' mix 85
    drying times *see* drying hempcrete
    'standard' mix 85
    variations in the binder 79-84
    variations in the hemp shiv 79
Hemp-LimeConstruct 10, 38, 39, 67, 73,
    77, 111, 169, 189, 207, 225, 249, 269,
    354
Hemp Technology Ltd 26-7

hempcrete
    'carbon sink' 57, 69, 120, 222
    characteristics 7, 62, 63-4, 65
        carbon sequestration 7, 9, 38, 56,
            57, 69, 97
        deformability 68, 90, 91, 137, 290
        elasticity 89-90, 137, 155
        fire resistance 7, 92-3
        good thermal mass 65, 68, 279, 311
        hygroscopicity 23-4, 65, 96, 99, 272,
            293
        insulation value 137, 221-30
        low-tech nature 10, 11, 29, 67, 71,
            209, 341
        monolithic structure 36, 64, 65, 97,
            119, 131, 142, 290, 292, 310, 320,
            322, 341
        non-load-bearing material 27-8, 68,
            90, 91, 155, 215, 290
        open matrix structure 94, 137
        porosity 65, 94, 97, 98
        vapour permeability 23-4, 26, 69, 70,
            71, 96, 293
    defined 23
    history of use 23, 25, 70, 347
        future challenges and possibilities
            347-50
    lifespan 69
    nature of the material 292-301
    performance *see* performance of
        hempcrete
    structural limits of the material 290-2,
        340
Heraklith 183
heritage buildings
    finishes 248, 272, 274, *280*
    floors 221
    indoor air quality 270-1
    infill panels 272-3, *273,* 274-5, *276,* 296
    insulation, retrofitting 39, 68-9, 268,
        269-71, *269,* 274, 278
        hempcrete floor slab 281, *281*
        lath-and-plastered walls 278
        solid-wall insulation *116,* 278-80,
            279, 280
    lime, use of 37, 41, *44,* 48, *50*
    natural building materials 267, *268*
    restoration and retrofit 24, 28, 58, 68-9,
        221, 267-8, 347
    shuttering 187-9, *188,* 273
    specialist skills requirement 270
    wall thickness 292
    wattle and daub, hempcrete
        replacement of 277-8, *277*
    see also historic buildings; vernacular
        buildings
hessian 107

hiring tools and equipment 103, 106, 197,
    355
historic buildings 23, 24, 48, 68-9, 187-9,
    201, 211, 267, 269, 270, 279, 292, 296,
    318
    see also heritage buildings
holistic view of a building 120, 270, 271,
    334
hop-up platforms 261
hoses 106
hot-dip galvanizing 299
    nails 104, 136, 161, *161,* 275
humidity regulation 7, 35, 39, 42, 56, 65,
    93, 94, 96, 97, 99, 131, 225, 249, 293
hung-tile cladding 250, 272, 274, 278,
    308, 333
hydrated limes *see* air limes
hydraulic limes 43, 44-6, *45,* 50, 51, 80, 98,
    192, 263
    artificial hydraulic limes 47
    feebly hydraulic limes 45, 50, 51, 236,
        242
    formulated hydraulic limes 45, 47
    natural cements 45-6
    natural hydraulic limes (NHL) 45, 50
    strengths 45
    vapour permeability 44, 46
hydraulic set 44, 45, 46, 47, 50, 51, 80
hydrophobic natural cements 46
hydrophobic renders 293, 294, 335-6
hygroscopicity 39, 42, 56, 65, 71, 267
    binders 93-4
    gypsum plasters 49
    hemp-fibre quilt insulation 38
    hempcrete 23-4, 65, 96, 99, 272, 293
    lime 42, 69
    shiv 93
hygrothermal behaviour 88, 96-7, 293

I-joists 158, *163, 328,* 342, 344
impact drivers *104,* 105, 174
indoor air quality 48, 55, 56, 65, 69, 94,
    98-9, 131, 249, 271, 293
    airtightness and 311
industry standards 25, 27, 76, 77, 83, 88,
    99, 289-90, 348
inhaling binder or lime powder 114
initial set 70, 79, 80, 139, 180, 211, 215, 264
insect attack, inhibition of 26, 68, 137
insulation 23, 24, 32, 62-3, 120
    additional 140-1
    airtightness at junctions 39, 297-9, *297,*
        311-12, 317, *325, 326, 326*
    ceilings 225-7, 330, *330*
    floors 221-4
    hemp-fibre quilt *see* hemp-fibre quilt
        insulation

hemp–lime mix and 85
inside the airtight line 310
lime–hemp plasters and 37
natural-fibre insulations 38, 56, 62, 327
plinths 64, 153, 312-15, *313*, *314*
pre-cast panels 33
retrofitting (heritage buildings) 268,
    269-71, *269*, 274, 278
    lath-and-plastered walls 278
    solid-wall insulation *116*, 278-80,
        279, *280*
roofs 85, 224-5, 227-8, *228*, 322-3,
    326-9, *329*, 330, *330*
standard thickness 131
synthetic insulation 38, 39, 54-5, 56, 65,
    119, 221, 268, 280
U-values 33, 34, 38, 62-3, 96, 140, 222,
    239, 291, 292, 319, 327
wood fibre 56, 62, 124, 140, 280, 298,
    324, *324*, 325, *325*, 326
see also thermal performance
internal walls 33, *130*, 133-4, 235, 249-50,
    292, 311, 321, 338
Isochanvre 93

Japan 24
jigsaws 103, 105
joiners 253-4, 255
joist hangers *162*, *163*, 299, 301
joists 158, 161-2, *163*, 164, *165*, 226, 272,
    301, 303, *326*, 328, *328*, *330*, 338

kitchens, finishes 249-50

labour-intensive nature of building with
    hempcrete 71, 121, 133
    see also workforce
labourers 254, 255
large-scale commercial builds 29, 30, 33,
    34, 122, 197
    see also build practicalities
laser levels 103
latex gloves 107, 112, 210
'leaky' buildings 55, 65, 160, 270, 271, 274,
    280, 296, 310, 311
Lhoist 27, 80, 83
lifecycle analysis (LCA) 36, 51, 54, 55,
    97, 183
lighting 107
lime 41-51
    advantages over cement 48
    air limes 42-4, 45, 46, 50, 51, 80-1, 93-4,
        215, 236, 263
    anti-rotting agent 68, 215, 293
    caustic nature of 98
        see also lime burns
    characteristics 50

dry hydrate of lime 42-3, 47
embodied energy 48, 68, 69
formulated hydraulic limes 45, 47
'free' lime 44, 45, 46, 236
hydraulic limes *see* hydraulic limes
hygroscopicity 42, 69
inhibition of insect attack 26, 68, 137
milk of lime 43
pozzolanic additives 46-7, 51, 80, 151,
    263
UK Building Limes Forum 41, 354
vapour permeability 32, 42, 44, 46, 49
lime-based binder 9, 23, 24, 26, 51, 63, 68,
    69, 70, *78*, 79, 80, 92, 98, 109, 123, 156,
    192, 194, 213, 263, *292*, 293, 302
    health-and-safety issues 109-10, 111
    storage and handling of bags 110-11
    temperature parameters 123
lime burns 110, 112, 113, 191, 262
lime cycle 42
lime–hemp plasters 37-8, *37*, 274
lime mortars 23, *31*, 41, 42, 43, *45*, 47, 48,
    59, 93, 135, *146*, 151, 152, 224, 246, *246*,
    251, 255, 277, 278, 306, *313*, *316*, 317,
    318, 319, 334, *339*
lime plasters *40*, *49*, 70, 72, *127*, 135, 231,
    232, 235-6, *236*
    addition of bast fibres and shiv 25, 37
    application techniques *see* plastering
    hygroscopicity 42
    lime putty plaster *42*, 43, 195, 263, 270
    lime–hemp plasters 37-8, *38*, 274
    'rougher' finish 232, *232*
    shrinkage 37
    staining 74-5, *74*, 216, 217
    standard two-coat finish 141
lime renders 43, *61*, 70, 72, *125*, 135, *141*,
    208, 231, *233*, 235-6, 245, *249*, 250, 251,
    272, 278, *280*, 295, 297, *306*, *317*, *318*,
    335, 336, *337*, *339*, 341
    lime render–timber cladding
        combination *250*, 251
    shrinkage 245, 295
lime screeds 251, 319
Lime Technology Ltd 27, 34, 83, 219
lime water 43
limecrete 149, 224, *316*, 318
Limecrete Company 29, 354
limewash 43, 51, 70, 74, 135, 247-8, 249,
    255, *280*
linseed-oil mastic sealant 245, 298, 312,
    317
lintels *59*, 60, 162, *162*, 163-4, *163*, 164,
    185
    design detailing 338
    engineered timber lintels 164
    structural lintels *37*, 164, 338

lithium-ion (Li-ion) tools 104
loft insulation 38
loose-fill cellulose insulation 314, 328
low-tech nature of construction 10, 11, 29,
    67, 71, 209, 341

magnesium silicate board 183, 184, *184*,
    321
masking tape 107
masks 110, 111, *111*, 196
masonry construction tools 107
mastic sealants 245, *245*, 246, 295, 298,
    312, 317, 326, *337*, 338, *339*
mechanical behaviour of hempcrete 89-90
mechanical ventilation and heat-recovery
    (MVHR) systems 55, 126, 311, 344
medieval timber-frame buildings 23, *24*
    see also heritage buildings
metal mesh, expandable 153
milk of lime 43
mixers 28, 105, 197, 355
    bell mixers *100*, 105, 106, 201-5, *204*
    pan mixers 105, *105*, *190*, *193*, 195-6,
        195, *198*, 201
    siting 197
mixing hempcrete 10, 28, 68, *70*, *71*, 72,
    140, 191-206
    basic principles 191-3, 199
    daily quantity of mix 201
    desired consistency 192
    health-and-safety issues 111
    in heritage properties 273
    instructions, adhering to 85, 195
    measuring using buckets 202
    method 196-9
    mixing and ferrying teams 254, 255-6
    mixing area 259
    mixing process 199-200
    mixing–ferrying–placing balance 201,
        205
    personal protective equipment (PPE)
        111
    proportions, accurate 70, 84-5, 192,
        193, 195
    testing the mix 191, 200, *200*
    tools and equipment 194-6
    water quantities 26, 70, 74, 140, 192-3,
        197, 262
        too little 74, 192
        too much 74, 192-3, 208
    weather conditions and 197, 262
    see also hemp–lime mix
moisture ingress 44, 94-5, 135, 137, 151,
    231, 245, 250, 273, 291, 293-6, 304,
    331, 334
    water penetration tests 94, 291
moisturisers 112, 113, 211

Moores, Bob and Tally 124, 126, *127*
mortars 45
    cement mortar 43, 152, 192
    hydraulic lime and sand 31
    lime 23, *31*, 41, 42, 43, *45*, 47, 48, *59*,
        93, 135, *146*, 151, 152, 224, 246, *246*,
        251, 255, 277, 278, 306, *313*, *316*,
        317, *318*, 319, 334, *339*
    structural performance 47
mould growth 27, 64, 65, 75, 76, 94, 99,
    267
multi-tools, powered 106, *106*

nail floats 106, *107*, 185, 188, *216*, 336
nail guns 104, 105, 299
natural cements 23, 45-6, *46*, 47, 50, 51,
    80, 151, 152, 215
    hydrophobic 46
    see also Prompt Natural Cement
natural hydraulic limes (NHL) 45, 263
natural materials
    breathability 58, 135
    construction industry adoption of 25,
        264
    hygroscopicity 56, 215
    negative net carbon emissions 57
    non-sustainable 56
    reasons for choosing 9, 56, 124
    vernacular and historic buildings 118,
        267, 274, 305
nickel-cadmium (NiCd) battery-powered
    tools 104
nickel-metal hydride (NiMH) battery-
    powered tools 104
noggins 91, *162*, *163*, 164, *165*, 303-4, *303*,
    314, 327, *328*, *337*, *339*
non-hydraulic limes *see* air limes
non-vapour-permeable materials 49, 58,
    60, 62, 221, 267, 277, 315

off-gassing 65, 99, 135, 221
openings
    airtightness around 338
    design detailing 336-8
    shuttering *139*, 140, 163, 176, 185-6,
        186, *187*, 338
    see also door openings; window
        openings
OSB (oriented strand board) 103, 104,
    138, 158, 160, 189, 201, 226, 259, 296,
    320
    frames 160, 296
    racking boards 320-1
    shuttering boards 138, 170, *170*, 171,
        173, 175, 177, 189, 226
overhangs 133, *133*, 166, 279, 293, 306, *306*

painters 255
paints 247-9
    clay paints 248
    making your own 248
    silicate paints 248
    toxic chemicals 247, 248
    vapour-permeable 70, 135, 247, *247*,
        248
    waterproof 249
pan mixers 105, *105*, *190*, *193*, 195-6, *195*,
    198, 201
panels, pre-cast 33-5, *33*, *34*, 51, 68, 132,
    356
Parex 335
performance of hempcrete 87-99
    acoustic performance 33, 88, 89, 91,
        97-8, 99, 126, 131, 133, 225, 311, 330
    ageing, effects of 91
    fire resistance 7, 92-3
    indoor air quality 48, 55, 56, 65, 69, 94,
        98-9, 131, 249, 271, 293
    mechanical behaviour 89-90
    research 87-9
    resistance to damage by moisture 94-5
    structural qualities 90-2
    thermal performance 7, 9, 26, 36, 54,
        62-5, 91, 94, 95-7, 119, 126, 284, 293
    vapour permeability and
        hygroscopicity 23-4, 26, 65, 69, 70,
        71, 93-4, 96, 99, 272, 293
    workmanship and 99
personal protective equipment (PPE) 109,
    111-12, 191, 210
pipework 142-3, *143*
pitched roofs 165, *165*, 225, *306*, 307,
    327-8, *328*
place names 16
placing hempcrete 28-9, *28*, 35-6, 69, 71,
    132, 133, 137, 140, 166, 207-19, *210*
    application tools 105-6
    around horizontal frame sections 213
    awkward placing 303
    basic principles 207-9
    compaction of placed material 69, 95,
        98, 99, *150*, 193-4, 207-8, 209, 211,
        214, 226, 274
    desired surface texture 208-9, *209*, 211
    eaves 302-3, *302*
    errors 214
    health-and-safety issues 111
    labour-intensive system 133
    lifts, bonding between 213
    localized areas of low-density
        hempcrete, dealing with 214
    method 209-14
    over-filling and cutting back 211, *211*
    personal protective equipment (PPE) 112

    placing team 209-10, 214, 254, *254*,
        256-7
    rate of placing 171
    regular review of 213-14
    safe working practices 210-11
    spray-applying 29-30, *29*, *30*, 69, 132,
        269-70, 296, 347
    test panels 209, 211, 213
    unfilled voids 214, *214*
    weather conditions and 262-3
planning permission 344
planning the build 117-23
    contingency planning 123
    costs 119-20
    design process 120
    inspiration 117-19
    practicalities 121-3
        drying times *see* drying times
        workforce factors 121
    see also build practicalities
plasterboards, gypsum 49, 133, 134, 183
plasterers 255
plasterer's baths 107
plasterer's hawks 236, *238*, *239*
plasterer's whisks 106
plastering
    alkali-resistant glass-fibre mesh 236,
        242, 243-4, *243*
    commencing 35
    'learning wall' 238
    on to hempcrete 236-41
    on to wood wool board 242
    personal protective equipment (PPE) 112
    skills 142, 255
    technique 233, 238-9, *238*
        basecoat application 238-9, 242
        corner reinforcement 244
        floating the surface 238
        rubbing back 42, 239, 240, *240*
        suction, managing 242
        topcoat application 239-40, 242
        wetting down *241*, 242
    tools and equipment 106-7, 236-7,
        237, *238*
    training courses 72, 232
    wall preparation 238
    weather conditions and 123
    working temperature range 236
plastering beads 244-6
    corner beads 244, 295
    sealing beads 245-6, *245*
plasters
    clay 135, 141, 234-5, *235*, 249, *278*
    colouring 233
    extra strength, adding 233
    gypsum plasters 48-9, *49*, 58, 72, 142,
        232, 244, 278, 293

lime *see* lime plasters
    mix ratios 237
    mixing method 237-8
    mixing your own 233, 234
    proprietary products 233, 236
plastic disc washers 184
plastic shuttering 189, *189*
plinths 133, 136, 147, 151-2, 160
    airtightness at 317
    brick 151, *152*, *317*
    cold bridging 153
    concrete 151-2, 153
    damp-proof course *see* damp-proof
        course
    design detailing 312-17
    designing out 315, *316*
    drip details 246, *246*, 295
    functions 312
    gabions 152
    insulation 64, 153, 312-15, *313*, *314*
        avoiding need for 314, *315*
    interaction with door openings 317
    low-embodied-energy plinths 151-2
    moisture ingress, prevention of 295-6
    multi-functioning materials 312-13
    non-structural plinth 315, *316*
    rammed-earth car tyres 152
    site-specific effects on 306, 307
    stone and lime mortar *146*, 151
    width 151
plumbing 142, 143, 173
plywood 160, 187, 296, 320, 336
polymer binder 56
Portland cement 23, 44, 45, 47-8, *48*, 80,
    94, 152, 215, 262
pozzolans 46-7, 51, 80, 151, 263
pre-cast hempcrete 30-5, 36, 347
    and cast-in-situ hempcrete compared
        32, 35-6
    blocks *see* blocks
    cost effectiveness 132
    costs 36
    embodied energy 36
    panels *see* panels
    pros and cons 30-1, 35
principles of hempcrete building 67-77
    the build 70-1
    nature of the material 67-70
    problems 73-7
    workforce 71-3
problems 11, 70, 73-7
    applying finishes too early 36, 74-5
    contractor errors 73, 74-5
    faulty materials 75-7
    incorrect amount of water in mixes 74,
        192-3, 208
    remedial work 102

slow drying 73, 74, 122, 264
wall slump and collapse 26, 69, 76, 77,
    77, 80, 192, 193, 222
processing plants 21, 26-7, 36
Prompt Natural Cement 46, *46*, 80, 81, 82,
    144, 171, 180, 211-12, 215, 223, 263,
    291, 296, 297, 355
protective clothing and equipment 106,
    109, 111-12, 191, 210
pry bars 105
pure lime *see* air limes

quicklime 42, 43

racking strength 90, 91, 155, 292, 319-21
    diagonal stainless steel straps 92, 164-5,
        321, 322, *322*
    diagonal timber bracing 92, 158, 164,
        321, 322, *322*
    OSB or plywood boards 320-1
    vapour-permeable plaster carrier
        boards 321
raft foundations 136, 148, *148*, 150, 151,
    301, *314*
rafters 39, 166, 172, *172*, 226, 227, 228,
    228, 268, 293, 301, *301*, 323, *324*, 325,
    325, 326-7, 329, 330, 332, *332*, 338
    bolt-on rafter ends 166, 323, *324*
    rafter hangers 301
rain, driving 262, 265, 291, 305, 306, 307
rain-screen cladding 124, 218, 251, 294,
    332
rammed earth 37, 64, 152
recycled glass foam aggregate 222, 224, 318
recycled glass foam blocks 153, *153*, 312-13,
    313, *318*
render stop beads *297*, 298-9, 312, 317,
    326, 338
rendering
    personal protective equipment (PPE)
        112
    skills 142
    technique *see* plastering
    tools 106-7
    weather conditions and 123
renders 311, 331, 332
    addition of shiv 37
    application techniques 233
    applying too early 216, *216*
    basecoat 239, *241*
    breathable 141
    cement render 58, 62, 72, 142, 221, 244,
        255, 267, 277, *317*
    creating your own 233
    hydrophobic renders 293, 294, 335-6
    lime-based 43, *61*, 70, 72, *125*, 135, *141*,
        208, 231, *233*, 235-6, 245, *249*, 250,

    251, 272, 278, *280*, 295, 297, *306*,
        317, *318*, 335, 336, *337*, *339*, 341
    lime–sand render 294
    proprietary products 233, 234
    rain, exposure to 306
    topcoat 240
Renewable House Programme (UK) 25,
    321
research 87-9
resources 354-6
resting moisture content 71, 193, 215
restoration and retrofit *see* heritage
    buildings
retardants 45, 46, 212
retting 21
reveals *127*, *139*, 185, *211*, *244*, *288*, 336,
    337, 338
rising damp 59, 60, 62, 221, 313
roofs
    airtightness 322, 323, 324-6, 327, 328,
        329
    design detailing 322-9
    eaves *see* eaves
    flat roofs 165, 225, 326, 329, *329*
    hempcrete roof insulation 85, 90,
        224-5, 227-8, *228*, 229, 322-3, 326-9,
        329, 330, *330*
        flat roof 329, *329*
        pitched roof 327-8, *328*
    overhangs 133, *133*, 166, 279, 293, 306,
        306
    pitched roofs 165, *165*, 225, *306*, 307,
        327-8, *328*
    prefabricated trusses 165
    rafters *see* rafters
    site-specific effects on 306-7
    vapour-permeable roof membrane 323,
        324, *325*, 326, *328*, 329
'rope walks' 17, *17*
rotting 26, 68, 135, 192, 215, 216, 259, 264

St Astier 80
sarking boards 323, 324-6, *324*, *325*, 327,
    328
saws
    circular saws 103, *103*, 104, 105
    guide rails 104
    hand saws 105
    hole saws 105
    jigsaws 103, 105
    slide mitre saws *102*, 103, 105
scaffolding 260-1, *260*, *261*
screws
    extra-long decking screws 173, 180
    heavy-duty screws *162*, 174, *174*, 175
second-hand tools and equipment 103
seismic design 90

self-builders 10, 29, 30, 72, 73, 121, 133, 185, 232
  Agan Chy 124-7, *125*, *127*
  Bridge End Cottage 342-5, *343*, *345*
  foundations 149
  Hemp Lime House 282-5, *285*
  volunteer labour 73, 121
services 142-5, *142*
  electrical installations 144, *144*, *145*, 280, 300, 303
  gas supply pipes 142-3
  plumbing 142, 143, 173
  water supply and waste pipes 142-3, *143*
set squares 102, 105
shaping hempcrete 105, 106, 140, 185, 186, 188, *188*, 216, *216*, 275
sheep's wool 56, 62, 271
sheeting 107
shelves 186
shiplap cladding 251, 332, *332*
shiv 15-16, 23, 24, 25-6, *25*, 88
  acoustic performance, impact on 97-8
  addition to plasters and renders 37
  binder–shiv mix *see* hemp–lime mix
  characteristics
    capillary structure 93
    flexibility 89
    hygroscopicity 93
    porosity 62, 89, 93, 293
  cost 27
  dust 26, 27, 76-7, 79, 98, 110
  faulty 76-7
  fines 21, 26, 27, 76, 79, 233
  grades 29-30, 79
  importing 27
  industry standards 25
  length of pieces 26, 79
  particle size distribution 97, 98
  pre-approval testing (France) 84
  processing 25-6
  rotting 26, 68, 135, 192, 215, 216, 259, 264
  sourcing 26-7
  storage 26, 192, 258-9
  suppliers 26-7, 289, 356
  sustainable resource 56
  vapour permeability 93
  variations in 79
  water absorption 76-7, 192
shrinkage 37, 81, 143, 153, 245, 274, 296-9, 312, 317
shuttering 27, 28, 30, 169-89
  accuracy, importance of 102, 167
  boards
    certification 183
    cutting area 259
    desired characteristics 169

environmental impacts of manufacturing 183-4
    full boards 171, 177, 181
    magnesium silicate board 183, 184, *184*
    OSB 138, 170, *170*, 171, 173, 175, 177, 189, 226
    overlap panels 175
    permanent shuttering boards 183
    re-using 104, 138, 169
    wood wool board 183, *183*, 184, *185*, 187, 233, 236, 242, 244, 259, 321, 336, *337*
  on central frames 173-7, *174*, *176*
    striking shuttering from 180-1, *180*
  chipboard 169-70
  corners 177-80, 185
    external 177, *178*
    internal (Method 1) 178, *179*, 180
    internal (Method 2) 178, *179*, 180
  curved walls 187, *187*
  door and window openings *139*, 140, 163, 176, 185-6, *186*, *187*, 338
  on exposed frames 180
    fixing and striking shuttering 181-3, *182*
  filling *see* placing hempcrete
  fixings
    alkali-resistant 184
    plastic disc washers 184
    spacers 138, 166, *166*, 167, 173, *173*, 175, 181
    stainless steel fixings 170
    straps 175, *175*, 180, 182
  framework design, relation to 166-7
  health-and-safety issues 111
  heritage buildings 187-9, *188*, 273
  lifts 70, 139, 171-2, 177, 181, 189, 212, *212*, 213
  one-person job 182
  permanent 135, 138, 160, 164, 165, 183-5, *184*, *185*, 244, 338, 341
    finishes 233-4
    fixing 183-5, *183*
    internal 183
    labour-saving option 184-5
  personal protective equipment (PPE) 111
  plastic shuttering 189, *189*
  plumb and square, checking for 139, 172, 177
  screws 173-4, *173*
    bends and head damage 174
    extra-long decking screws 173, 180
    heavy-duty screws 174, *174*, 175
  shuttering team 255, 257, *257*
  speed of 171, 182, 183

temporary 70, 135, 138-9, 166-7, 169-83, 338
  basic principles 170-3
  boarding out one side of wall in advance 171
  filling tops of walls from above 172
  labour-heavy option 185
  'lollipops' 172, *172*
  method 173-83
  quantities of 139, 170-1
  striking 70, 71, 79, 80, 102, 139, 141, 171, 172-3, 180-1, *180*, 215
  tools 105
  two-person job 177, 181
silica 44-5
  see also magnesium silicate board
silicate paints 248
sills 340, *340*
site organization 258-61
site topography and climate 304-6
skin
  cuts 113, 114
  protection *see* barrier creams; moisturisers
slaking 42, 43, 45, 46
slates
  cladding 135, 250, 274, 278, 305, 306, 331, 336
  damp-proof layer 152
  drip detail 246, *246*
  sills *339*, 340, *340*
small ground movements, toleration of 68
Smith, Bill and Jo 342, 344, *345*
Society for the Protection of Ancient Buildings (SPAB) 268
sockets 144, 213, 280
softwood frames 28, 134-5, *134*, 136, 137, 138, 155, 157, *159*, *165*, 170, *180*, 315, 321, 331, *332*, 333, 341
solar energy 124, *125*, 126
sole plates 158, 160, *160*, 161, *297*
spacers 138, 166, *166*, 167, 173, *173*, 175, 181
  filling spacer holes 181, *181*
  removal *181*
spade bits 105
specialist consultants 120
spillages
  dust 111
  hempcrete 113, 171
spirit levels 105
sponge floats 107, *107*
spray-applying 29-30, *29*, *30*, 69, 132, 269-70, 296, 347
stainless-steel cladding ties 334, *334*
stainless-steel strapping 92, 164-5, 299, 321, 322, *322*

Stanley knives 106
steel 155, 156, *156*, *157*
 anti-corrosion coating 300, *300*
 corrosion 299, 300
 frames 33, 132, 136, 155, 156, *156*, *157*, 301, *301*, 315, 320
 galvanized steel 160, 299, 300
 replacing steel with timber fixings 300-1
 stainless steel straps 92, 164-5, 299, 321, 322, *322*
 stainless-steel cladding ties 334, *334*
 straps 160, *160*, 161, *162*, *163*, 299
stone cladding 255, 335
stonemasons 255
straight edge 107
straw-bale buildings 64, 250, 349
strip foundations 147-8, 149
structural engineers 92, 149, 152, 155, 319-20
structural frameworks *see* frames, structural
structural integrity 26, 32, 50, 85, 88, 192, 195, 208, 262, 274, 290, 291, 299, 320, 333, 338, 340
structural qualities of hempcrete 90-2
 ageing, effects of 91
 hempcrete combined with a timber frame 91-2
 hempcrete on its own 90-1
 testing protocol, desirability of a 92
studs 136, 138, 158, 159, 160-1, *162*, 164, 213, 272
 fixing 161, *161*
 function 160
 multiple or reinforced studs 161, 163, 164, *165*
 spacing 161
 wall falling away from 159
studwork frames *see* frames, structural
supervisors 253, 254, 255, 257
sustainable building: key concepts 53-65
 breathability 58-62
 indoor air quality *see* indoor air quality
 natural insulation 38, 56, 62, 327
 sustainable construction materials 55-7
 thermal performance *see* thermal performance
 zero carbon 7, 54, 57
Sustainable Traditional Buildings Alliance (STBA) 268, 354
synthetic insulation 38, 39, 54-5, 56, 65, 119, 221, 268, 280

tadelakt *187*, 249-50
tamping stick 207-8, *208*, 311
tamping team 254, *254*

tannins 74, 217
temperature regulation, internal 63, 64, 97
 see also indoor air quality
Tempo retardant 46, 212
thermal bridging *see* cold bridging
thermal conductivity 38, 64, 96
thermal envelope 33
thermal mass 24, 32, 33, 58, 63-4, *63*, 85, 119, 124, 270, 323
 hempcrete 63-4, 65, 68, 99, 279, 311
thermal performance 7, 9, 11, 26, 33, 36, 37, 38, 39, 54, 55, 56, 58, 62-5, 68, 69, 74, 88, 91, 94, 95-7, 119, 120, 126, 131, 211, 264, 267, 268, 271, 279, 280, 282, 284, 293, 304, 311, 313, 318, 344, 347
 historic buildings 68-9
 variables 88, 95-6, 131
 see also insulation
ties 159, 250, 320
tiled surfaces 249
timber
 cladding *125*, *141*, 250, *250*, 251, 255, 306, 331, 332, *332*, *333*
 diagonal timber bracing 92, 158, 164, 321, 322, *322*
 engineered timber joists 158, 165
 engineered timber lintels 164
 floors 39, 153, 224, 251, 271, *315*, 319
 gussets 300-1, *300*, 322, *322*
 hardwood 135, 138, 156, *156*, 159, 291, 315, 331, 333
 I-joists 158, *163*, *328*, 342, 344
 preservatives 134-5
 softwood 28, 134-5, *134*, 136, 137, 138, 155, 157, *159*, *165*, 170, *180*, 315, 321, 331, *332*, 333, 341
 storage 259
 trussed joists 158
Timber Research and Development Association (TRADA) 157-8, 354
timber-frame buildings 23, *57*, 91-2, 124, 320
 see also heritage buildings
Tomorrow's Garden City, Letchworth *57*, *141*
tongue-and-groove cladding 251
tool belts 105
tools and equipment 101-7
 basic toolkit 107
 checking for accuracy 102
 cost and quality 101-2
 essential and desirable tools 103-4
 frame and shuttering tools 105
 hand tools 102
 hempcrete application tools 105-6
 hiring 103, 106, 260, 355
 masonry construction tools 107

plastering and rendering tools 106-7
professional quality 103
resale value 103
second-hand 103
standard toolkit 105-7
storage 259
toxic emissions 7, 9, 55, 99, 221, 248
Tradical® HB 27, 80, 82, 83, 84, 219, 223, 292, 355
Tradical® Hemcrete® walling system 84, 99, 294
Tradical® HF 27
training courses 41, 72, 232, 234, 355
transportation costs 27, 36, 51, 152, 225
The Triangle, Swindon *11*, *233*
Troldtekt 183
trowels 107, *107*, 233
trussed joists 158
try squares 105

U-values 33, 34, 38, 62-3, 96, 140, 222, 239, 291, 292, 319, 327
UK
 binder suppliers 348
 British hemp industry 16-17, 18, 19-21, *20*
 Building Limes Forum 41, 43
 Building Regulations 63, 91, 140, 271, 291, 292, 319
 construction industry 25, 53, 264, 349
 hem shiv suppliers 26-7, 348
 LABC binder pre-approval testing 84
 Renewable House Programme 25, 321
 research centres 87, 321
 zero-carbon-homes target 54
underfloor heating 224, 251, *316*, *318*, 319, 344
USA 18-19, 347-8

vapour permeability 9, 33, 39, 50, 67, 88, 137, 183, 231, 232, 234, 235, 240, 242, 249, 251, 267, 271, 277, 281, 293, 295, 296, 315, 320, 323, 324, 326, 327, 329, 333, 335, 336, 338
 air limes 93-4
 binders 80, 93-4
 carrier boards 228, 233, 321
 ceilings 225-6
 cladding 250, 330, 331, 332, *332*, 334
 finishes 74, 141, 193, 216, 224
 floors 221, 222, 317-19
 hempcrete 23-4, 26, 69, 70, 71, 96, 293
 lime 37, 42, 44, 46, 49
 natural cements 80
 paints 70, 135, 247, *247*, 248
 shiv 93
 walls 58, *59-61*, 62, 71, 209, 215, 249, 264

vapour-permeable membranes 58, 135, 295, *295*, 296, 302-3, *302*, 331, 332, *332*, 333, 334, *334*, *339*
ventilation
mechanical heat-recovery ventilation systems 126, 311
natural 65, 311
vernacular buildings 118-19, *118*, 152, 267, 305, *305*
see also heritage buildings
Vicat 80, 81, 212, 291
see also Prompt Natural Cement
volatile organic compounds (VOCs) 55, 65, 225, 247
volunteer labour 73, 121

wall plates 158, 161, *162*, *163*, *297*, *301*
walls 131-45
cast-in-situ hempcrete 131-47
construction method 135-45
door and window openings 139-40, *139*
finishes 141-2
foundations 135-6
hempcrete 140-1
plinths 136
services 142-5, *142*
shuttering *see* shuttering
structural frame *see* frames, structural
construction principles 131-5
curved walls 187, *187*
desired surface texture 208-9, *209*, 211
drilling into walls 143, 144
drying *see* drying hempcrete
filling tops of walls from above 172
fixing into 137, *138*
internal 33, *130*, 133-4, 235, 249-50, 292, 311, 321, 338
thickness 133
rain, protection from during placing process 262, 265, *265*, 307
straightness 102, 139, *167*, 176
structural maintenance 135
thickness 133, 140, 291-2
site-specific effects on 306
vapour permeable 58, *59-61*, 62, 71, 209, 215, 249, 264
water ingress *see* moisture ingress
water penetration tests 94, 291
water supply and waste pipes 142-3, *143*
waterproof clothing and boots 113, 196, 262, *262*
wattle and daub 67, 268, 274
hempcrete replacement of 277-8, *277*
weather
build practicalities and 261-4

drying hempcrete and 35, 121, 123, 217-18, 307
mixing process and 197, 262
placing hempcrete and 262-3
plastering and 123
rendering and 123
safety precautions 113
site topography and climate 304-7
West Bay, Dorset 17
wet weather
driving rain 262, 265, 291, 305, 306, 307
safety precautions 113
wheat straw 56
whole-thermal-envelope design 120, 221, 310, 326-7
wicking action *48*, 222, 284, 313, 335
Wild, Leah 282, 284, *285*
window openings 139-40, *139*
airtightness around 338, 340
beads
designing out 245, *245*, 295, *337*
frame seal beads 245-6, *245*, 295, 338
hardwood beads 245, *245*, 295, *337*
design detailing 336-40
horizontal sections *337*
vertical sections *339*
drip details 246, *246*, 294-5, *294*, *295*, *334*, 340
fixing to structural frame 336
hemp quilt insulation *337*, *339*
lintels *see* lintels
moisture ingress, prevention of 294-5, *294*, *295*
placing hempcrete 211
plastering 243
reveals *127*, *139*, 185, 211, *244*, *288*, 336, *337*, 338
shuttering 185-6, *186*, *187*, 338
sills 340, *340*
timber around openings, protection of 295, *295*, 338
window framework 164
Womersley's 38
wood fibre 56, 62, 124, 140, 280, 298, 324, *324*, 325, *325*, 326
wood wool board 183, *183*, 184, *185*, 187, 233, 236, 242, 244, 259, 321, 336, *337*
wood-fibre sarking board 323, 324-6, *324*, *325*, 327, *328*
wooden straps 175, *175*
workforce 71-3
balancing the team numbers 255-7
ferrying team 254, 256
frame-construction team 253-4
mixing team 193-4, 254, 255-6
placing team 209-10, 214, 254, 256-7

shuttering team 255, 257, *257*
skills and experience 72-3, 121, 348-9
supervision 73, 121, 195
training 72, 83, 121, 195, 210, 214
volunteer labour 73, 121
workmanship
consistency of 29, 71
hempcrete performance, importance to 99
quality of 68, 71, 83

zero carbon 7, 54, 57